Story of the 1900
GALVESTON HURRICANE

GOVERNOR SAYERS, OF TEXAS.

Story of the 1900
GALVESTON HURRICANE

Edited by
Nathan C. Green

PELICAN PUBLISHING COMPANY
Gretna 2000

Copyright © 1900
By R. H. Woodward Company

First edition, 1900
First Pelican edition, 2000

The word "Pelican" and the depiction of a pelican are trademarks of Pelican Publishing Company, Inc., and are registered in the U.S. Patent and Trademark Office.

ISBN 1-56554-767-5

Printed in the United States of America

Published by Pelican Publishing Company, Inc.
1000 Burmaster Street, Gretna, Louisiana 70053

Mrs. Mary Moore, a survivor of the 1900 Galveston hurricane

PREFACE.

The *Story of the 1900 Galveston Hurricane* is a graphic account of one of the most remarkable disasters of modern times. The island on which the city stands was swept by the waters of the Gulf of Mexico for six hours, accompanied by a storm which lashed the whole Texas coast with terrible fury for sixteen hours, and even now the cause of the rising of the waters is but imperfectly understood. The wind blew, if the testimony of those present is to be believed, from the northeast, veering around to the east. Its force was from sixty to one hundred miles an hour, and it would seem that their force was such as to drive the waters out to sea faster than they could flow in. But the flood came, and the island was covered to the depth of three feet in the highest places, and there the waters remained, though lashed to foam by the force of the hurricane, a sort of inexplicable riddle of the apparent violation of the laws of nature.

This book deals with the storm and its terrible effects upon the people and property of Galveston. It was a casualty, or a series of casualties, for which it is not easy to find a parallel.

The catyclism in which so many lives in Johnstown were sacrificed comes readily to mind, but it was a visitation much shorter in duration and less terrible in its effects. It brought destruction swift and merciless to one small city, while the Texas hurricane devastated miles and miles of territory, destroyed many small villages, wrecked farms, and at Galveston wrought havoc such as finds no equal in the historical records of the New World. Some have read of the earthquake which visited one of the islands of the East Indies a few years ago and caused the destruction of almost all the inhabitants on the island. The victims of that tidal wave were numbered by the thousand, and it alone stands beside the awful work of the deluge and storm which wrecked the city of Galveston on the day and night of the 8th of September, 1900.

It has been the purpose of the writer in the preparation of this book to present as a vivid a picture of the storm as possible and to depict the horrors of those terrible days at the time of the storm and after in the language of those who were on the spot and saw with their own eyes what had taken place. Acknowledgments must be made to one of the gifted writers of the Galveston News, some of whose descriptions have been used entire. This writer was in his own home on the night of the storm, saw and experienced with the keen observation of a newspaper man trained to such work, and then wrote as no one could write who had not been in the thick of the storm and amid the heartrending scenes immediately afterwards,

It is hoped that the book will be found, what it was intended to be, a faithful and trustworthy history of one of the greatest casualties in history.

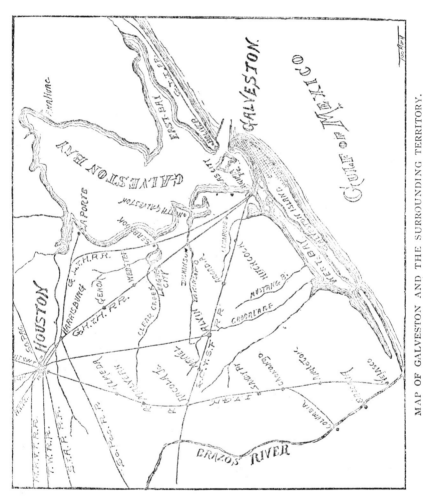

MAP OF GALVESTON AND THE SURROUNDING TERRITORY.

CONTENTS.

PART I.
STORY OF THE STORM.

CHAPTER I.
The Terrible Hurricane................................... 3

CHAPTER II.
The Storm Sweeps Inland............................... 25

CHAPTER III.
How the News Came..................................... 33

CHAPTER IV.
After the Storm... 46

CHAPTER V.
Survey of the Wreckage................................. 63

CHAPTER VI.
Newspaper Woman's Story............................... 85

CHAPTER VII.
The People in Despair................................... 102

CHAPTER VIII.
The Work of Relief...................................... 116

CHAPTER IX.
Galveston's Earnest Appeal............................. 144

CHAPTER X.
Under Martial Law....................................... 158

CONTENTS.

CHAPTER XI.
Galveston Redivivus.................................... 176

CHAPTER XII.
Storm Scientifically Considered........................... 201

CHAPTER XIII.
Galveston Had Warning................................. 224

CHAPTER XIV.
Homes on the Sand.................................... 241

PART II.
STORIES TOLD BY SURVIVORS.

CHAPTER I.
As Dr. Cline Saw It..................................... 251

CHAPTER II.
Newspaper Man's Narrative.............................. 258

CHAPTER III.
Father Kirwin's Story................................... 272

CHAPTER IV.
Strange Scenes in a Convent............................. 282

CHAPTER V.
Some Personal Experiences.............................. 287

CHAPTER VI.
Sights in Galveston..................................... 324

PART III.
Sketches of Galveston and Texas......................... 341

LIST OF ILLUSTRATIONS.

Governor Sayers of Texas	Frontispiece
Map of Galveston and the Surrounding Territory	x
At Pier 12	9
One of the Morgues	19
Cotton Exchange	20
Corner Avenue I and Twenty-second Street—Church on Corner Was Destroyed	20
Showing Strip Four Blocks Wide and a Half-Mile Long Entirely Denuded	29
Beach Front—Nothing Left	39
Pagoda Bathhouse—Nothing Left to Show Even Where It Stood	39
Near Sixteenth and M	49
From Pier 23, Looking East	59
Near L Street	69
Volunteers Removing Debris to Open Street, Under Guard, Twenty-first, Looking North	79
Corner Strand and Twenty-second Street, Looking West—One Story Blown Off Moody Bank Building	80
In Front of Catholic Church on Broadway	89
Galveston Wholesale Hat and Shoe Company, Four-Story Brick	99
Burning the Dead	100
Near Sixteenth and K	109
View of Galveston	119
No Mercy for Ghoulish Looters at Galveston	129
Sacred Heart Church (Jesuit), Thirteen Blocks from East Beach—Only One House Left	139
City Railroad Company Power-House, Looking West	139
Twenty-ninth and Market—Taylor Compress	140
Mallory Wharf, S. S. Alamo	140
Scene Near the Ocean Front	149
Texas City Wharf	159

LIST OF ILLUSTRATIONS.

Pier 21	159
What Is Left of the Residence Portion from the Ursuline Convent to the Beach, a Distance of Six Blocks	169
Refugee Camp, Furnished by the U. S. Government	179
St. Patrick's Church, Twenty-fourth and Avenue K—Had the Highest Steeple in the City	179
Thirty-third Street Pier	180
View from Fourteenth and K	180
Near Pier 16	189
Large House to Left Moved Thirty Yards Without Injury—Nineteenth and N½	199
Tremont and Avenue L, Looking South	199
View from Fifteenth and K	209
City Water Works and Electric Plant—Nearly Totally Destroyed	219
Elevator "A"	229
Burying the Dead on Spots Where Found	239
From Pier 33, Looking East	240
Sixteenth and M	249
"Kendall Castle" Blown Through Bay to Texas City	259
The House in the Foreground Was Lifted Off Its Blocks and Carried Fifteen Squares	259
Back of Grand Opera House	277
First Baptist Church, Corner Twenty-second and Avenue I, East Side	277
In the Business District—Big Retail Stores on Market Street	278
Debris Worked from Galveston Into Texas City	278
A Refugee Camp	295
One of Galveston's Principal School Buildings—Many Persons Were Killed in This Building	295
Back of Mallory Wharf	296
Near Elevator "A"	296
Elevator "B"	313
Near Eighteenth Street Wharf	313
Seeking the Wounded and Securing the Dead Amid the Ruins on Market Street	332
Galveston and Texas Coast Swept by an Awful Hurricane	349

Galveston News.

MONDAY, SEPTEMBER 10, 1900.

FIRST NEWS FROM OUTSIDE.

Relief Sent from Houston—Supplies Landed at Virginia Point.

Galveston, Monday P. M.

The citizens of Houston have responded to the needs of the city of Galveston, and are making an effort to do something for those in need here. Yesterday in Houston, Col. R. M. Johnston and Hon. John Henry Kirby, started a movement for relief supplies for Galveston, and the people of that city responded liberally.

The first of the relief party to arrive reached here at 1 o'clock to-day. It was a delegation consisting of James Hayes Quarles of the Post staff, Charles Quinn and C. H. Hartnell and V. R. Jarrell of the fire department. They left Houston at 4 o'clock this morning on a special train on the Galveston, Houston and Henderson in charge of Chief J. J. Hussey of the fire department. On the train were two hundred and fifty volunteers who came to bring assistance to injured and to bury the dead. The train brought a carload of provisions for Galveston. The train could come no closer than within five miles of the bay. The provisions have been carted to the bay shore and the gentlemen walked that distance and then came in a rowboat to this city to get boats to go after the provisions. They reported to Mayor Jones, and arrangements were started to get the boat.

In addition to the above the steamer Lawrence left Houston this morning in charge of a committee of citizens. It brings two carloads of provisions and four tanks of water containing 25,000 gallons each.

Mr. Quarles says that Houston suffered by the storm to a small extent. There was a large number of buildings damaged and some wrecked, but they were serious cases. But one life was lost, and that was a man who went against a live wire. Houston is in total darkness, the electric light plant being damaged. No streetcars are running.

The storm has extended to quite a number of towns in the interior, and there is some loss of life and damages. All of the bridges across the bay are wrecked, not a foot of rail being on the trestle.

Dead bodies are found all along the railroad track from Houston. The railroads have no track within four miles of the bay. When the gentlemen left Houston they had heard of no damage at New Orleans.

Reports reaching the State are that Galveston is wrecked and that 2,500 souls were lost, and one report said one-third of the population have gone.

ORGANIZING RELIEF.

First Meeting Held Sunday Afternoon. Committees Appointed.

Galveston, Sunday.

At 2 o'clock Sunday afternoon a meeting of prominent citizens was held at the Chamber of Commerce for the purpose of organizing to relieve the suffering and to bury the dead. The meeting was organized with Mayor Walter C. Jones as chairman and ex-Congressman Miles Crowley as secretary.

A motion was adopted that the chairappoint a central committee of nine, with the Mayor as chairman. This was adopted, and the Mayor announced as the other eight members, Messrs. B. Adoue, John Sealy, I. H. Kempner, Jens Moller, W. A. McVitie, Ben Levy, M. Lasker and Daniel Ripley.

The meeting was then adjourned and the following members of the committee, being all that were present and could be reached, met: Mayor Jones and Messrs. Sealy, Moller, Lasker, Ripley and McVitie.

On motion the chairman appointed chairmen of committees on finance, correspondence, hospital, burials, general relief. The committees were at once filled out and got to work. The burial committee is charged with collecting and burying the bodies of all dead human beings and animals.

Chairman Levy of the burial committee suggested that it would greatly delay matters to hold inquests on all the dead bodies. Mayor Jones said he would call a meeting of the City Council to pass an ordinance suspending the provisions requiring inquests.

Clarke & Courts advised the committee that they would advance all the money immediately necessary.

RELIEF COMMITTEE.

A meeting of the relief committee was held on Sunday evening. Mr. W. A. McVitie, chairman, presiding. W. C. Ogilvy was elected secretary of the meeting.

On motion of Mr Forster Rose the chairman was empowered to employ a permanent secretary at a salary by the week.

The chairman suggested that the city be districted, that the roll of the committee be called, and a chairman be elected for each ward. This was done, and the following ward chairmen, to have charge of the relief work in their respective districts, were elected.

First ward—Mr. Thomas Doyle.
Second ward—Mr. Charles L. Wallis.
Third ward—Mr. Jake Davis.
Fourth ward—Mr. A. C. Torbett.
Fifth ward—Mr. C. H. McMaster.
Sixth ward—Mr. George Stenzel.
Seventh ward—Mr. Forster Rose.
Eighth ward—Mr. Edmond Bourke.
Ninth ward—Mr. Clarence Ousley.
Tenth ward—Mr. W. F. Coakley.
Eleventh ward—Mr. John Goggan.
Twelfth ward—Mr. Edgar J. Berry.

The relief committees for the different wards were empowered and instructed to give immediate relief wherever necessary.

The following have been appointed by Chairman McVitie as purchasing committee: Charles Willis, Jake Davis, M. Ullmann, G. H. Mensing, Gus Levy and H. C. Lang; C. H. Dorsey has been appointed secretary.

SUB-COMMITTEES.

The Work of Bringing Order Out of Chaos Begun.

Galveston, Sunday.

The general committee appointed chairmen of the several sub-committees, and each chairman appointed his own assistants. These sub-committees are as follows:

Correspondence: M. Lasker, chairman; Col. R. G. Lowe, Mayor Walter Jones, Clarence Ousley, J. D. Skinner, and C. H. McMaster. This represents the press, the city, the Cotton Exchange and the Chamber of Commerce.

Finance: John Sealy, chairman; M. Lasker, B. Adoue, I. H. Kempner, W. L. Moody Jr., J. Moller, L. B, Bergeron.

Hospital: Daniel Ripley, chairman; Rabbi Henry Cohen, Father Kirwin, Dr. Wm. Scott, Dr. W. F. Starley and Dr. H. A. West.

Burial Committee—Ben Levy, chairman; J. Stoner, F. P. Habine, Frank Sommers, Sterling Norman, M. F. Wirt.

General Relief—W. A. McVitie, chairman; C. H. McMaster, H. Mosle, J. H. Hawley, Paul Jones, W. B. Wallis, Joe Gengler, Chas. Wallis, Jake Davis, M. Ullmann, Will Mensing, Willie Lyle, Gus Lewy; Edmund Bourke, W. C. Ogilvy, Rabbi Cohen, M. McLemore, Wm. Hanscom, W. F. Morrisey, Forster Rose, Tom Doyle, Max Levy, Ald. Webber, Joe Scott, M. M. Mann, J. Wharton Terry.

J. H. Hawley was appointed to chairmanship of his own committee to see that property was protected from thieves and depredators.

At Mr. Moller's suggestion the relief committee was authorized to appoint a secretary to stay at some specified place from 7 a. m. to 7 p. m. to hear petitions.

Alderman Levy, chairman of burial committee, said that the law required a coroner to sit upon each dead body and that to expedite burying the 400 to 600 dead it was necessary to act at once. Mayor Jones said he would, at once, call a meeting of the city council and pass an emergency measure obviating this necessity.

Mr. Moller, on the part of the finance committee, proposed a plan to get money from the United States government.

Mr. Bourke, of Clarke & Courts, said that the money would advance all the money necessary.

It was stated that the Gulf, Colorado and Santa Fe would be called upon at once or cash.

MONDAY'S WORK.

Able-Bodied Men Who Refuse to Work Will Not Be Fed.

Galveston, Monday.

A second meeting of the relief committee was held Monday morning, with Chairman W. A. McVitie presiding, and Messrs. W. C. Ogilvy and Charles L. Dorsey as secretaries.

The question of removing debris from the streets was discussed at considerable length. It was reported that it was difficult to get men to work at clearing away the debris. A great deal of looting has been going on, and others who have not indulged in this practice have gotten supplies from the relief committee without working.

Mr. J. H. Hawley spoke pretty warmly upon this matter, and said no able-bodied man who refuses to work should be given food.

Chief of Police Ketchum, who arrived on the scene at this moment, said that the committee should seize all food supplies in the city at once—should notify every wholesale grocery and the flour mills that all the food they had belonged to the committee and would be paid for by it. Then the committee should take charge of the distribution of these supplies, and should permit no able-bodied man to eat unless he worked.

Mr. Hawley said the city should be placed under martial law and all food supplies guarded.

A messenger was dispatched to notify the wholesale grocers and the flour mills that the committee had taken charge of all their food supplies.

On motion of Mr. Forster Rose it was ordered that the chairman of each ward should have charge of the removal of debris in his ward, and should have charge of the food supplies for that ward, giving orders to no able-bodied men for food unless they work.

Volunteers for assistant secretaries were called for and Messrs. W. N. Fritter and Ross responded.

Chief of Police Ketchum also announced that Mr. Tom McHenry had taken charge of the work of clearing the streets so that ambulances and wagons could pass.

Over a hundred men are at work uncovering the machinery of the water works, and every effort is being made to get the water supply turned on again.

Headquarters of Edmund Bourke, chairman Eighth ward relief committee, are at Garten Verein bowling alley.

The steamship Comal of the Mallory line arrived from New York this morning and is lying off the Mallory pier unable to attend her cargo.

Dead bodies are still being brought into the morgues. The work of burying dead, humans and animals, is progressing much faster to-day than it did yesterday, as there is now some organization and system to the work.

Instruction to Chairman of the Various Ward Relief Committees.

Galveston, Sept. 10.

It is your first duty to provide a central storehouse, as near to your ward as can be reached by a dray.

Second. Order by dray load from any wholesale grocery such staples as flour, sugar, coffee, crackers, meal, canned meats, salt bacon, tea, potatoes, etc. It's your duty to have assistants, who, together with yourself, will remain from 7 a. m. to 6 p. m., and supply all the people in your ward direct. You must not issue orders for small quantities on any wholesale store, but feed them from your central point.

Third. It is your duty to organize a working force, with horses and drays if possible, and clean up the debris in your ward in such manner as you deem best.

Any able-bodied man who will not volunteer for this work must not be fed.

Fourth. The Central Committee will meet at Goggan's building, corner Twenty-second and Market streets, at 9 a. m. daily, to confer and report.

Supply of Water.

It is hoped to get the waterworks running by to-morrow morning. In the meantime no one need suffer for water, as the stream is running from Alta Loma to the receiving tank.

FACSIMILE OF GALVESTON NEWS, MONDAY, SEPTEMBER 10, SECOND MORNING AFTER THE STORM.

Story of the 1900
GALVESTON HURRICANE

PART I.

STORY OF THE STORM.

"Story of the 1900 Galveston Hurricane."

CHAPTER I.

THE TERRIBLE HURRICANE.

THE CITY OF GALVESTON SWEPT BY THE WIND AND BURIED IN THE WATERS OF THE GULF.

STORM RAGES SIXTEEN HOURS.

THE PEOPLE IN TERROR LEAVE THEIR HOMES TO BE DROWNED IN THE STREETS—FLYING TIMBERS AND FALLING WALLS.

"One of the most awful tragedies of modern times has visited Galveston. The city is in ruins, and the dead will number possibly 6,000. The wreck of Galveston was brought about by a tempest so terrible that no words can adequately describe its intensity, and by a flood which turned the city into a raging sea. The wharf front is entirely gone, every ocean steamer in the harbor stranded, with a money loss that cannot now be estimated, is, so far as can be learned at this hour, a résumé of the appalling calamity that has befallen Galveston. The great storm has left her helpless, and her stricken people are compelled to appeal to the outside world for aid. The estimates of loss of life vary between the figures given, but an accurate count of the dead is impossible now, and the real number killed in the storm will probably never be known. No one attempts to estimate the damage to business and residence property."

This terse telegram was one of the first which conveyed to the outside world the terrible calamity which fell upon the city of Galveston on Saturday, September 8, 1900. In the graphic style of the press reporter, the correspondent went on to say that the water-works are in ruins and the cisterns were all blown away, so that the lack of water is one of the most serious of the present troubles. Ruin is everywhere. Electric-light and telegraph poles are nearly all prostrated, and the streets are littered with timbers, slate, glass and every conceivable character of debris. There is hardly a habitable house in the city, and nearly every business house is badly damaged. The school buildings are unroofed, such edifices as the Ball High School and Rosenberg School buildings being badly wrecked. The fine churches are almost in ruin. The elevators and warehouses are unfit for use, the electric-light plant has collapsed, and so has the cotton factory.

From Tremont to P streets, thence to the beach, not a vestige of a residence is to be seen. In the business section of the city the water is from three to ten feet deep in stores, and stocks of all kinds, including foodstuffs, are total losses. The correspondent in Galveston Saturday night saw women and children emerging from once comfortable and happy homes dazed and bleeding from wounds, the women wading neck-deep, with babies in their arms. To add, if possible, to the calamity, the city is cut off entirely from the world. The telegraph lines are down and the cable which connects Galveston with Mexico is cut. Nothing is to be seen of any of the bridges which connected the island with the mainland, but where the bridge should be a big ocean vessel is stranded.

Very few, if any, buildings escaped injury. There is hardly a habitable dry house in the city.

When the people who had escaped death went out at daylight to view the work of the tempest and the floods they saw the most horrible sights imaginable. In the three blocks from Avenue N to Avenue P in Tremont street I saw eight bodies. Four corpses were in one yard.

The whole of the beach front for three blocks in from the Gulf was stripped of every vestige of habitation. The dwellings, the pavilions, the great bathing establishments, Olympia and every structure having been either carried out to sea or its ruins piled in a pyramid far into the town, according to the vagaries of the tempest.

The first hurried glance over the city showed that some of the largest structures, supposed to be most substantially built, suffered the greatest.

At Texas City the wharves are destroyed, and the water front for a mile is littered with ruins, much of the debris having been blown there from Galveston. At Texas City three lives were lost. The railway track is washed away, and the only exit was by foot and conveyance to La Marque, on the International & Great Northern Railroad.

Saturday, September 8, dawned upon the Island City bright and clear, and her people were a happy, prosperous and contented community. Before 9 A. M. Galveston boasted proud supremacy as the commercial center of the Southwest, but 9 A. M. witnessed the beginning of the end —the downfall of a city's greatness.

The storm began at 2 o'clock Saturday morning. Previous to that a great storm had been raging in the Gulf, and the tide was very high. The wind at first came from the north and was in direct opposition to the force from the Gulf. While the storm in the Gulf piled the water upon the beach side of the city, the north wind piled the water from the bay on to the bay part of the city.

From 9 o'clock on fitful, uneasy gusts of wind swept over the fated metropolis. The waves in the Gulf and bay grew choppy, and they were snappy and menacing. Showers of rain swept along the streets, but no one thought of approaching danger, except, perhaps, a few of the more timid strangers in the city, who always did fear, and "knew that Galveston would sooner or later be blown away." They were not exactly frightened, but they were decidedly uneasy, and these strangers wished they were at their homes on the mainland.

By 9.30 in the morning there was a fairly stiff little gale blowing. Those natives of the city accustomed to the petty blows which come up at times on the Mexican Gulf coast rather admired the scene, and watched the picturesque breakers dashing upon each other and sending aloft fountains of dainty white foam. To them it was an inspiring and altogether enchanting spectacle. By 10 o'clock the severity of the wind had increased considerably. The water was several blocks inland on both the bay and Gulf sides and was rapidly rising. Then a number began to feel decidedly alarmed, and as the gale increased in fury, and water rose from the Gulf on one side and bay on the other, those residents located near the shore became panic-stricken. They looked out of their doors and found the waters creeping upon their galleries. Many of them, realizing what was near, left their homes and started out to wade, and in some cases actually swam to the business portion of the city.

By 11 o'clock every part of the city was covered with water from two to four feet deep, and the storm was increasing in fury. Houses near the beach, the majority of which are built on stilts five to ten feet above ground, commenced to waver and give way. The little bathhouses on the sands were swept out to sea, and with an ominous roar the big Pagoda Bathhouse, two squares long, fell suddenly into the angry waves. It was a crash that was heard for many squares, giving fearful warning to the people that catastrophe impended.

About noon it became evident that the city was going to be visited with disaster. Hundreds of residences along the beach front were hurriedly abandoned, the families fleeing to dwellings in higher portions of the city. Every home was opened to the refugees, black or white. The wind was rising constantly and it rained in torrents. The wind was so fierce that the rain cut like a knife. By 3 o'clock the waters of the bay and Gulf met and by dark the entire city was submerged. The flooding of the electric-light plant and the gas company's factory left the city in darkness.

From this time on the real part of that awful drama began. Fiercer and fiercer blew the wind; higher and higher rose the water. Consternation was everywhere, and even the old settlers who had been through the storm and flood of '75 became frightened and commenced precautions for abandoning their homes to seek shelter in the more substantial buildings in the downtown portion of the city. By 12 o'clock the waters from the Gulf had inundated the island as far as Twelfth street. The wind at that time was blowing eighty miles an hour, when the indicator was blown away and the weather station was wrecked. Rain was falling in torrents, and in the lower parts of the city men were hastening to places of safety, and women and children were struggling from waist to neck deep in the rising waters, which, blown in from the Gulf, covered the streets and were fast inundating the city.

To go upon the streets was to court death. The wind was then at cyclonic height, roofs, cisterns, portions of buildings, telegraph poles and walls were falling and the noise of the winds and the crashing of the buildings were terrifying in the extreme.

Some of the women bore children in their arms, and

AT PIER 12.

stumbled along in the blinding storm, not knowing which way to turn. Many men who had left their families early in the morning to go to their places of business, when they realized what danger their families were in, started homeward to protect the lives of their wives and children, but even while they were struggling to their loved ones, their wives and children were making frantic efforts to make their way to the business portion of the city, where they felt they would be safe.

The mecca of the refugees seemed to be the Tremont Hotel, as that building was known to be more than substantial. It was the shelter of many hundreds during the storm and flood. This hostelry fast filled up, and the influx of terror-stricken citizens, mostly men, continued until the refugees were packed on the upper floors with scarcely breathing space. The school buildings, courthouse, Union Depot and other substantial buildings were crowded to overflowing with terror-stricken men, women and children, huddled together promiscuously, seeking refuge from the pitiless storm which, before another day, was destined to sweep into eternity between 5000 and 8000 souls and desolate the greatest shipping and commercial center of the Southwest.

From Twelfth street the waters gradually encroached farther inland, rising about fifteen inches an hour. At 6 P. M. there were thirty-six inches of water in the lobbies of the Tremont Hotel, the highest point in the city. Across the street, where the ground is lower, a horse was drowned. At 9 o'clock the water on Market street was level with the seats of the street cars. After that it gradually receded, but the wind was of hurricane force, reaching a velocity of eighty-four miles an hour. In the streets the wires were down, telegraph and telephone poles falling, and slates and glass and timber were flying through the air.

The Tremont Hotel is the highest point on the island, and before 9.30 the water in the lobby was over the desk and covered the pages of the register. Across the street the sidewalks were low, and many were drowned in sight of those who had sought refuge on the second floors of the hotel. About the same time the Politz, a new three-story building just completed, collapsed; then came the caving in of the big Ritter Building, in which three of the city's most prominent citizens lost their lives, and one after another could be heard the crack of collapsing houses.

The wind and the waters rose steadily from dark until 1.45 o'clock Sunday morning.

During all this time the 40,000 people of Galveston were like rats in traps. The highest portion of the city was four or five feet under water, while in the great majority of cases the streets were submerged to a depth of ten feet.

To leave a house was to drown. To remain was to court death in the wreckage.

Such a night of agony was possibly never equalled by people in modern times.

Brave men started out wading and swimming in an endeavor to save life, but they were unable to reach much of the residence section of the city, and had to give up the attempt in order to save their own lives. Many of them were drowned or killed by flying debris before they were able to return to the safety which they had chivalrously deserted. The horror of the calamity grew with every passing moment. The air was filled with flying slates, bricks and glass from broken window panes. The real tragedy had commenced. From out of the doomed buildings poured maimed and bleeding men, women and children. Women were wading waist-deep in water with their dead children in their arms; men, dazed and in half-frenzy, searched frantically for their families. The moans of those dying and the shrieking of the living were frightful sounds as they mingled with the roar of the gale and the surging of the waters. The electric-light plant collapsed and the stricken city was in total darkness. This added to the horror of the situation, as no one knew what terrible calamity would follow next.

At times people would sail rapidly by in boats, and, colliding with some obstruction, would be painfully injured. Dr. S. O. Young, secretary of the college, was driven from home. He seized a board, and was whirled with terrific velocity toward the bay. Striking some obstruction, he was severely cut and bruised about the head and face, besides receiving bodily injuries. Dr. West, one of the prominent physicians of Galveston, was drowned near the Rosenberg School building, whither he had gone to attend a patient who was reported to be injured.

Without apparent reason the waters suddenly began to subside at 1.45 A. M. Within twenty minutes they had gone down two feet, and before daylight the streets were practically freed of the flood waters. In the meantime the wind had veered to the southeast.

Not until Sunday dawned did anyone know of the stupendous catastrophe. All knew their own experiences, but little did any of those who survived dream of the awe-inspiring results of the tempest and its effect on the living and the dead. By daylight the wind had subsided, and the water had run down until it was easy to wade about the stricken city. People who had taken refuge uptown or in stauncher houses than their own started with the first streak of daylight to look for some trace of the missing members of their families. Intuitively they sought places where their own homes stood. What scenes greeted them! Everywhere fallen houses, ruin and every conceivable kind of wreckage. Houses were blown over, unroofed and many demolished. Desolation was everywhere. Bruised and bleeding, and with anguish of despair, men and women walked the streets. Nearly all had lost everything on earth except the clothing on their backs. Occasionally a dead body would be seen. It caused a shudder to creep over the searchers.

When those who had escaped the fury of the storm reached the Gulf beach side of the city a terrible scene was presented. For five blocks the entire shore front had been swept of everything. Not a single house stood, and on Sixth street the wreckage had formed a sea wall thirty to forty feet high. Fully one-third of the houses had been destroyed, and not a house was standing that was not damaged 50 per cent.

A few of the cooler heads called a meeting, and two newspaper men were dispatched by boat to the mainland, thence to Houston with telegrams, informing the outside world of the disaster and asking for aid. This was done before even a cursory investigation had been made, and the loss of life was placed at between 800 and 1000, the latter figure being then thought excessive. Immediately upon receipt of the news, Houston organized a relief society and sent a boat loaded with provisions to the stricken city.

On Monday, after the storm had subsided and all the water had receded, the gruesome result of the devastation was revealed. People almost nude walked the streets appealing in vain for their lost relatives and friends. Some were bereft of reason, others were stoical, but none the less sorrowful. It was a most pitiful scene to witness. On all sides there was ruin, and on every turn dead bodies were encountered. In some places as many as forty and fifty corpses were huddled together. Screaming women, bruised and bleeding, some of them bearing the lifeless forms of children in their arms; men, broken-hearted and sobbing, bewailing the loss of their wives and children; streets filled with floating rubbish, among which there were many bodies of the victims of the storm, constituted part of the scene.

The first loss of life reported was that at Reuter's saloon on the Strand, where three persons lost their lives, and where many others were maimed and imprisoned. The dead were Stanley E. Spencer, Charles Kilmer and Richard Lord. These three were sitting at a table on the first floor when suddenly the roof caved in, killing all of them instantly. Those in the lower part of the building escaped with their lives in a miraculous manner. The falling roof and flooring were caught on the bar, the people standing near it dodging under the debris. It required several hours of hard work to get them out. The negro waiter, who was sent for the doctor, was drowned at the corner of the Strand and Twenty-first street and his body was found a short time after. The next place visited was the City Hall. Here were congregated fully 700 people, most of whom were more or less injured. One man from Lucas Terrace reported the loss of fifty lives in the building from which he escaped. He himself was severely injured about the head.

On Avenue M several ladies were imprisoned in a residence by the water and debris. They were rescued by a party headed by Capt. M. Theriot. Several of them were badly hurt. Coming back to Tremont street, and going out to Avenue P by climbing over the piles of lumber which had once been residences, the rescuing party observed four bodies in one yard and seven in one room in another place, while as many as sixty bodies were to be seen lying singly and in groups in the space of one block. A majority of the bodies, however, that had not been recovered were under the ruined houses, and it took several days' hard work to get all of them out. The body of Miss Sarah Summers was found near her home on the corner of Tremont and Avenue F, her hands grasping her diamonds tightly. The report from St. Mary's Infirmary showed that only eight persons escaped from the hospital. The number of patients and nurses could not be ascertained, but the number of inmates was seldom under 100. Rosenberg Schoolhouse, which was chosen as a place of refuge by the people of that locality, collapsed. Some of those who had taken refuge there escaped, but many were crushed to death in the ruins.

As Sunday morning dawned the streets were lined with wounded, half-clad people, seeking the aid of physicians for themselves and for friends and relatives who could not move. Police Officer John Bowie was found in a pitiable condition. He reported that his house, with wife and children, had been swept into the Gulf. The beach resort of Pat O'Keefe, who is known to every visitor to Galveston, was annihilated, not a vestige of the building remaining. Mrs. O'Keefe was drowned. The great bathing pavilion known as the Pagoda, the big pleasure resort known as the Olympia, and Murdoch's bathhouse were all swept away into the Gulf. There were a few bodies on the beach. They had been swept into the Gulf or driven up into the rubbish by the waves. Only half a dozen of them were in sight from the site where the workers were. All the residences which escaped destruction were turned into hospitals, as were the leading hotels. There was scarcely one of the houses which were left standing which did not contain one or more of the dead as well as many injured.

The rain began to pour down in torrents, and the rescuing party went back down Tremont street toward the city. The rain added greatly to the general distress. Stopping at a small grocery store, the party found it packed with the injured, clamoring for food, but the provisions in the store had been ruined. Further down the street a restaurant, which had been submerged by water, was serving out soggy crackers and cheese to the hungry crowd. On returning to the Tremont Hotel the correspondent found the death list to be swelling rapidly, the accounts coming from every portion of the city. Information from both the extreme eastern and extreme western portions of the city was difficult to obtain, but the reports which were received indicated that those two sections had suffered fully as much as the rest of the city. Fifteen men, constituting all that remained of a company of regular soldiers stationed at the beach barracks, were marched down Market street. The loss of life among these soldiers in the barracks, which were destroyed, must have been fully 100. At 11.30 Sunday morning the water had receded from the higher portions of the city, but the streets near the bay front still contained from two and one-half to three feet of water.

The Galveston News office, on Mechanic street, was flooded and the back end of the building caved in. At the Union Depot were scenes similar to those met with in other portions of the city. Baggagemaster Harding picked up the lifeless form of a baby girl within a few feet of the station. Its parents could not be located, and are supposed to have been lost. The station building had been selected as a place of refuge by a large number of people. All windows in the building and a portion of the wall at the top were

ONE OF THE MORGUES.

COTTON EXCHANGE.

CORNER AVENUE I AND TWENTY-SECOND STREET—CHURCH ON CORNER WAS DESTROYED.

blown in. The water around the station was probably twelve feet deep.

On the water front the destruction of property was almost as great as on the beach, though the loss of life was not nearly so large. The wharves of the Mallory Company were completely destroyed. The big steamship Alamo was lying among the ruins of the pier. The wharves of the Galveston Wharf Co. were also gone, and the great wharves of the Southern Pacific Company, which had been in couıse of construction for several months, were damaged to the amount of $60,000. The Norwegian steamship Gila, engaged in the Cuban trade, was stranded up the bay beyond where the railroad bridge once stood. The British steamship Taunton was lying on Pelican Island hard aground. The Mexican, a big British steamer, was driven up the bay and was fast in the mud. Another big ship was lying out near Quarantine Station. The Kendal Castle was driven as far up as Texas City, where she stranded. Of the small shipping only a few boats were left. Many of the little schooners were lifted bodily out of the water and flung up on the island. The wrecks of others were scattered along the bay front. The Charlotte M. Allen, the steam ferry-boat to Bolivar, was safe. The big dredge used at Texas City was driven inland for half a mile and could not be saved. The Pensacola was in port when the storm began, but Master Simmons put to sea in the teeth of the storm and the boat and her crew of thirty-six men were lost.

The three grain elevators and Rymerschoffer mill were

wrecks. Their roofs and the top stories had gone, and grain stored therein had been ruined by the rain. The damage to the ships at the time when the demand for tonnage was great was regarded as one of the worst features of the disaster from a business standpoint.

In the business portion of the city the loss could not even approximately be estimated. The wholesale houses along the Strand had about seven feet of water on their ground floors, and window panes and glass protectors of all kinds were demolished. The top of the Moody Bank building was blown away, and the fixtures of every house on this long business thoroughfare were destroyed. On Mechanic street the water was almost as deep as on the Strand. All provisions in the wholesale groceries and goods on the lower floors were saturated and rendered valueless. The engine-house of the Tremont Hotel was caved in by the falling smokestack. The damage to the hotel building amounted to $25,000. The power-house of the street railway company was destroyed, and the loss on machinery and building was roughly estimated at $70,000. There were no wires of any sort standing. They were lying in tangled masses across the streets, and had to be cleared away before horses and vehicles could move about the streets. All the bathhouses on the beach were destroyed and their attendants drowned. The Sealy Hospital was destroyed and most of the patients were drowned. The grain elevators were destroyed, one of them containing 1,000,000 bushels of wheat. The Ball High School and the Rosenberg School buildings were destroyed, and many persons who had taken refuge in them were killed. Eight big steamships in port were wrecked. All three railroad bridges and the county bridge across to the mainland at Virginia Point were swept away, and the bridge-tenders and their families drowned. The entire island was submerged, and water was eight feet deep on Tremont avenue, probably the highest point in the city.

So far as describing the scene of wreckage is concerned, it was out of the question. On the East End all residences and small stores were leveled. It looked like a lumber-yard over which had been strewn furniture. There were a number of bodies in there yet, for the work of rescue was necessarily slow. In the West End, where most of the poor people lived, and where houses were not so substantial, the damage amounted almost to a totality. There were a large number of dead there also.

In the wealthy part of the city there was one house out of five gone completely, or so badly wrecked that it was not habitable. All the houses left standing were severely damaged.

Eight ocean steamers were torn from their moorings and stranded in the bay. The Kendal Castle was carried over the flats of the Thirty-third Street Wharf to Texas City and lay in the wreckage of the Inman pier. The Norwegian steamer Gyller was stranded between Texas City and Virginia Point. An ocean liner was swirled around through the west bay, crashed through the bay bridge, and was lying in a few feet of water near the wreckage of the railroad bridges.

The steamship Taunton was carried across Pelican Point, and was stranded about ten miles up the east bay. The Mallory steamer Alamo was torn from her wharf and dashed upon the Pelican flats and against the bow of the British steamer Red Cross, which had previously been hurled there. The stern of the Alamo was stove in and the bow of the Red Cross was crushed.

Down the channel to the jetties two other ocean steamships lay grounded. Some schooners, barges and smaller craft were strewn bottom side up along the slips of the piers. The tug Louise of the Houston Direct Navigation Co. was also a wreck. The shores at Texas City contained enough wreckage to rebuild a city. Eight persons who were swept across the bay during the storm were picked up there alive. Five corpses were also picked up. There were three fatalities in Texas City. In addition to the living and the dead which the storm cast up at Texas City, caskets and coffins from one of the cemeteries at Galveston were being fished out of the water.

Along the bay shore the devastation was complete, and the full force of the wind could be better appreciated when the great timbers and pilings were seen twisted and broken in two. Many of the residences that were blown inside out had a look of frailness about them, but there was a solidity about the wreckage that showed that terrible force must have been exerted to have caused the damage.

CHAPTER II.

THE STORM SWEEPS INLAND.

A PATH THROUGH TEXAS TWO HUNDRED MILES WIDE DEVASTATED BY RAIN AND WIND.

SIXTY-FIVE TOWNS WRECKED.

GREAT EXTENT OF COUNTRY DEVASTATED, PEOPLE KILLED, STOCK DESTROYED, CROPS RUINED AND BUILDINGS BLOWN TO ATOMS.

Had it not been for the unparalleled destruction at Galveston the hurricane would have been regarded as a fearful storm from the havoc wrought outside of that city on the Gulf coast and further inland. From Red river on the north to the Gulf on the south and throughout the central part of the State, Texas was storm-swept for thirty hours by a hurricane which laid waste property, caused large loss of life and effectually blocked all telegraphic and telephonic communication. The operation of trains was also seriously handicapped. Starting with the hurricane which visited Galveston and the Gulf coast on Saturday, the storm made rapid inroads into the center of the State, stopping long enough at Houston to damage over half of the buildings of that city. Advancing inland, the storm swept into Hempstead, fifty miles above Houston; thence to Chapel Hill, twenty miles further; thence to Brenham, thirty miles further, wrecking all three towns.

The Brazos bottom, which was the scene of such disaster last year as the result of the flood, suffered a large share of damage by the hurricane, and was swept for fully 100 miles of its length, everything being turned topsy-turvy by the high winds, and much destruction resulting to crops, as well as farmhouse property. The winds were accompanied by a heavy rainfall, which served to add to the horror of midnight.

The storm swept all the tableland clear of everything on it, razing houses to the ground and tearing trees up by the roots. It also swept into the mountain gorges, and there inflicted the worst damage, and considerable loss of life was reported from that section as soon as communication was secured. From Southwest Texas and points along the Gulf coast to the city of Galveston the reports were alarming, and particularly concerning Galveston and Rockport. A number of parties summering at various points along the coast were heard from. The cotton had been nearly ruined, as the storm swept the cotton belt, and reports from all available sections were to the effect that the crop has been swept as clean of its fruit as though by the hand of man, and will be almost a total loss.

Owing to the excessive rains this year the cotton had grown to weed more than ever known, and in some fields it ranged from six to ten feet high, and was very rank with leaf.

The wind whipped and twisted the stalks, beating the open cotton out of the burr, and the rain beat it into the ground, so that it was ruined. That which was not yet open was all that was left, and the boll weevil had been doing damage to the young cotton, so that the outlook was gloomy, so far as this section is concerned.

At Richmond, Texas, many buildings were totally destroyed. The courthouse was greatly damaged. The Baptist church was a total wreck, while the Methodist church was almost destroyed. Three lives were lost in the colored Baptist church—Henry Ransom and two children, colored. One person was killed at Booth, and four lives were lost at Beasly.

After the storm Letitia was a wreck. The houses which stood in the place, including a depot, were blown to the ground and the timbers from some of them carried for miles. Mrs. Sophia Schultz, formerly of Houston, was killed.

On Sunday after the storm the International & Great Northern Railroad ran out two trains in the hope of getting through to Galveston. The second train left at 5 o'clock and got as far as La Marque. It remained there till midnight and returned, as the track beyond was washed away and it was impossible to reach Virginia Point, whence boats cross to Galveston. All along the route were evidences of the ruin and havoc wrought by the elements. From one-half to two-thirds of the telegraph wires were down, poles being broken off short.

At Genoa the depot building was a total wreck, and all

the other buildings were severely twisted, some being demolished. At Webster the wind had completely demolished half the houses in the town, and one house was turned completely bottom side up, resting on its gable. At Cedar Creek stores and residences had been blown down, and at Dickinson the havoc was almost complete. Six persons were killed and a number of dead cattle lined the side of the road, having been killed by flying timber. The prairie was covered with water, and tubs, buckets, doors and timbers had been carried several miles by the wind.

Refugees from Virginia Point said everything there was swept away, and that the beach was strewn with pianos and household effects from Galveston. One man saw six dead bodies on the beach. At Virginia Point the houses were all destroyed, and the members of the crew of the relief train reported that four or five colored people were drowned there, and two children of a Mr. Wright perished. At Dickinson the buildings were blown away, and several fatalities were reported.

A relief train got as far as Texas City. The Inman Compress, a very large plant of the kind, was a complete wreck. The whole beach front was lined with wreckage from Galveston. Four bodies had floated over from Galveston, and five persons had been rescued at that point. The British steamer Kendal Castle was ashore and proved a total wreck. H. F. Matthews, from five miles north of Virginia Point, reported the death of ten persons at Dickinson. At Texas City Junction, he said, the station agent's family of five were

SHOWING STRIP FOUR BLOCKS WIDE AND A HALF MILE LONG ENTIRELY DENUDED.

drowned. A score of people were ashore near Virginia Point in a helpless condition, he reported, but most of these were brought up by the next relief train.

At Eagle Lake three churches, together with many houses, dwelling and business, were completely blown to pieces in Saturday night's storm. Crops of all kinds were almost ruined. The loss to that community was estimated at about $250,000. No lives were lost there, but the town of East Barnard was blown away and three persons were killed there. There were seventy-five villages and towns that were swept by the storm, and from most of these places loss of life was reported. It was estimated that the loss of life, exclusive of the death list of the island of Galveston and the city of Galveston, was at least 600. Towns were wiped out of existence and many deaths were reported. The storm was over 200 miles wide and extended as far inland as Temple, a distance of over 200 miles from the Gulf. The velocity of the wind throughout this territory was greater than ever before experienced. It was not a steady blow, but it came in gusts, and seemed to have a cyclonic movement, which wrenched roofs off brick buildings and twisted trees from their trunks. There would be a severe gust of wind lasting a few seconds, and then a lull for a few seconds. This kept up for about twenty-four hours, and few buildings, no matter how substantially constructed, were able to withstand it. The cotton crop in the lower counties in Texas was completely ruined. The same is true of the rice crop.

The distress was keenly felt by the planters and small

farmers throughout the storm-swept region. In Brazoria and other counties of that section there was hardly a plantation building left standing. All fences were also gone, and the devastation seemed complete. Many large and expensive sugar refineries were wrecked. The negro cabins were blown down and many negroes were killed. On a plantation a short distance from the town of Angleton three families of negroes were killed. The death list of that town alone numbered fifteen persons. Thomas Forton of Fort Bend county reported that the villages of Needville and Beasly in that county had been completely destroyed and over twenty people killed, most of the bodies having already been recovered when he left. Every house in that part of the country was destroyed, and there was great suffering among the homeless people. Bay City suffered a loss of nearly all of its buildings, and three were killed there. There were many homeless people in Missouri City. Every house in the town except two was destroyed. People were living out of doors and camping on the wet ground.

No such storm had ever visited Texas, spreading such devastation and so much suffering in its wake.

CHAPTER III.

HOW THE NEWS CAME.

THE FIGHT MADE BY THE DALLAS NEWS SATURDAY NIGHT FOR THE DETAILS OF THE DISASTER.

EVERY KNOWN MEANS USED.

TELEGRAPH AND TELEPHONE COMPANIES, SPECIAL MESSENGERS AND RAILROADS EMPLOYED.

RECEIPT OF FIRST DREAD TIDINGS.

Never before in all the years of its history has this paper experienced so much difficulty in furnishing its patrons with the news.

Heretofore enterprise, backed up by money, has enabled it to be successful, but in this instance, although both had been tried, neither had been successful.

Saturday night a messenger boy in blue rushed into the office bearing a message containing a bulletin of this, undoubedly the greatest, disaster the Lone Star State has ever known.

Quickly the night editor turned to the operator on the News' special wire to Galveston:

"See if you can get Galveston," he said.

The operator had just reported for duty. Without taking off his coat or hat he tried to work his instrument.

"Can't do it," he replied at length; "the Western Union says that all South Texas wires are down."

"Get Houston, then."

A few more ticks, and then the same reply.

"Well, Austin or San Antonio."

And still the same reply.

"Here," exclaimed the managing editor, "work the Mexican cable and Vera Cruz, and get New Orleans and Atlanta ends to the story."

In a few minutes every means at the disposal of modern science through its marvelous inventions was being employed to furnish this paper with the news—all of it, if possible, but, in any event, at least an inkling.

"We can give you a story via Vera Cruz," was the message that soon came.

It was instantly ordered.

That it was meager and unsatisfactory readers of the News know, but it was the only word received from the ill-fated Island City that night.

More was ordered.

"Give us every line you can get," were the instructions.

But in a few seconds came the announcement: "The cable is gone; it has parted somewhere near Galveston."

New Orleans and Atlanta were bare of information.

Houston and Galveston were shut off from the world, and, for all anyone knew, had been wiped out of existence.

Telegram after telegram was sent to the correspondents of the News in South Texas.

Soon a hundred trained and alert newspaper men were straining every nerve to get even one detail, no matter how unimportant.

The telegraph and telephone companies gladly rendered all the assistance in their power.

The local staff was given instructions to glean from every source all the information possible.

But the hours went by, and nothing came.

The public is familiar with the story published yesterday morning. It is not familiar with the struggle against the elements made by the newspaper men of this State.

Half a thousand times yesterday did the telephone bell in the local rooms of the News ring.

Each time the question heard was: "Have you heard anything from Galveston?" and each time was a negative reply made.

Beginning early in the morning, the News again employed every instrumentality to hear from the scene of the catastrophe.

Telegraph and telephone companies, the railroads and special messengers were called into requisition.

Late in the afternoon communication was established with Houston.

That city had been visited by a terrible cyclone. Property had been destroyed, trains delayed, and it was feared that there had been great loss of life. The electric plant had been demolished and the city was in darkness.

As dusk was beginning to creep up over Dallas two trains arrived in the city. One was from Houston, and came over the Houston & Texas Central, and the other came over the Santa Fe from Milano Junction. The passengers thereon brought reports which confirmed those already received.

But none knew of Galveston's fate. None could tell whether one life or a score of lives, whether half a dozen structures or all in the city had been lost.

Houston was again appealed to, and the word came back that every device known to man would be used to establish communication with the place fifty miles south.

To this was appended a dispatch telling of the almost total annihilation of nearby towns.

It was just 9.15 P. M. when Mr. James Ward, the Dallas representative of the Associated Press, gave an exclamation and handed out a dispatch.

It was dated Houston and was in the form of an interview with a man who had just escaped from Galveston.

One thousand human beings, he asserted, had lost their lives and 4000 buildings had been destroyed.

Later came another dispatch placing the number of casualties at near 2000.

A relief train, which had been unable to cross the bay, returned to Houston.

Its passengers said they had counted 200 corpses scattered about in the lowlands.

The first authentic news from the doomed city was carried by a few passengers who left the Grand Central Depot shortly after daylight on Sunday morning and afterwards reached Houston and other points, and told of an awful night of storm and disaster.

On Sunday afternoon a correspondent of the Houston Post worked his way across the bay, and by degrees to Houston, and wrote the first story of the disaster as he had seen it, a portion of which appears in the opening chapter of this book. It was not complete, and did not give much of the details of the terrible havoc wrought by water and wind, but it told to the world the fact that a great city, and the chief emporium of the Southwest, had suffered the worst of any city of modern times. It set forth the needs of the destitute people and unfastened the purse-strings of the world to minister to their help.

Gradually detail to the awful picture was added, and the people of the United States knew of the most terrible catastrophe that had ever visited their land. It could only be compared with the ravages of fever in Philadelphia in the closing years of the last century, with the Chicago fire, the Johnstown horror and some of the simoons which have from time to time visited the shores of India and the islands of the sea.

In the manner described the news was brought from Galveston by the few refugees who could plod through the wrecked country to the railroad and thus reach their destination.

The press work in the main was well and conscientiously done. Reporters of papers and of the press associations began to reach Galveston by Tuesday morning, and truthful stories were sent out as fast as telegraph facilities would allow. Of these the people of Galveston had no reason to complain.

But with the conscientious reporters came those of the yellow journals of New York and Chicago, and of these they did justly complain.

On this point a correspondent of the Dallas News wrote to his paper:

Galveston people have experienced yellow journalism at long range, but they have recently had it brought home to them with a vengeance.

A few days ago a former New Yorker said: "The yellow journals of New York do a great deal of good; they also do a great deal of harm." He spoke rightly. The people of Galveston greatly appreciate what the New York papers have done in the way of raising money for suffering Galveston and in sending relief here for those who need it. But they have no excuse for those of the correspondents who sent out lying reports, some of whom would have permanently injured the city and sent it to its death if they could.

They are willing to overlook inaccuracies, which were unavoidable in the confusion of the first few days following the storm, and pardon the use of rumors of loss concerning certain matters before the exact conditions could be learned. But the New York correspondents did not reach here until

BEACH FRONT—NOTHING LEFT.

PAGODA BATH-HOUSE—NOTHING LEFT TO SHOW EVEN WHERE IT STOOD.

after the worst was over, and they might have had the truth if they would and had possessed one-tenth of the pluck, energy and astuteness of which they have boasted. But some of them not only did not search for the truth, but they absolutely failed to print the truth concerning things which they saw with their own eyes.

God knows the storm and its consequences were horrible enough, and if the yellow-journal correspondents had hustled for the news and had gotten the truth, and had possessed a vocabulary broad enough to express even faintly what they saw, they would have had stories thrilling enough to suit the most morbid appetite.

The conduct of some of these correspondents would be considered amusing had their product not turned out to be so damnably vile and so devilishly mendacious. They hugged the Tremont Hotel like they were afraid of venturing out into the districts where bodies were being cremated and men were working at the point of a bayonet, and constructed their stories on the statement of any old person, adding to that such trimmings as a vivid imagination and morally perverted heart could conjure up.

The people of Galveston had supposed that it would be impossible to exaggerate the wreck and ruin of the storm or its consequences, but it is evident that we have had a few things to learn concerning the writers of sensational stories, who care only for a text to work upon and the amount per column to be received.

A New York paper of Wednesday, September 12, contained a report from its special correspondent dated "Galveston, Texas, Monday, 9 A. M., by tug Juno to Houston, Monday night." That correspondent did not arrive in Galveston until Tuesday morning, yet he caused that paper to print, with all the sanctity of a statement from a special correspondent who was on the ground and had seen with his own eyes, several things which were at absolute variance with the facts. Somehow this correspondent slipped a cog, and in relating his tough experience in reaching Galveston, made the statement that he did not arrive in Galveston until Tuesday morning. And yet in the dispatch supposed to have been sent from Galveston by him on the day before he speaks with great pomposity of what he saw. In speaking of the robbery of bodies, he added to the horror of the story by saying that even the dying were robbed, and killed if they resisted—a thing not heard of up to date. He also said that negroes, business and professional men were driven to the water front and forced to handle dead bodies at the point of a bayonet. There was no forcing of business and professional men. They handled the dead bodies voluntarily, and begged and pleaded with negroes standing by to assist, and as enough men did not come to their assistance, they invoked the aid of the military government. He stated that "a dozen cases of assaults on white women have been reported at headquarters this morning." No such report was ever made. Then follows the nonsensical statement that the tugs Juno and Lawrence brought 2000 gallons of water from Houston. To this he adds that "half the population is without any water," that the town is suffering from a water famine, that "nobody washes his hands or face, and a bath is unheard of, and that the city is threatened with an epidemic as well as famine." All of which were lies of the purest ray serene.

The correspondent then attacks Galveston's port facilities. The wharves of the Galveston Wharf Co., he said, were almost blown to sea and almost all of its tracks destroyed. The Southern Pacific terminals, he said, were among the first to be wrecked, and the prophecy of the directors of the company, who opposed Huntington's hobby, had been realized by the blowing of the Southern Pacific wharves into ribbons. To this he adds that no Galvestonian will deny that the hurricane has sealed Galveston's doom as a shipping port. The correspondent had no authority for this statement, and had he rode into Galveston on a tug, as he claimed to do, and had he known a wharf from a side of bacon, he could have seen that the wharves had not been swept to sea, and that he landed upon one of them.

But if the work of the correspondent was yellow before he reached Galveston, it continued that way after he had been here for four days. Mr. I. H. Kempner, in yesterday's issue of the News, took care of his untruthful statement concerning the action and condition of the Galveston banks. Some of the statements made in the same dispatch were fair and of a hopeful tone, but are immediately negatived by slanderous and unfounded statements.

The Herald got some wonderful pictures for that Sunday paper. Under a headline running across the page they were printed. The headline read: "Photographs of the Ruin Wrought in Galveston by the Great Storm. The picture on the left shows the remains of the power-house of the Galveston Street Railway Co., where forty men were killed by falling brick and timbers. On the right the Mallory Line steamship Alamo is shown as she was cast up on the docks by the rush of the water, and also some of the wreckage swept up by the angry sea."

Governor Sayers' picture is in the center of the page, and is authentic. The alleged picture of the wreck of the city railroad power-house does not in the least resemble the ruins. Water is shown all around them. A News reporter passed that place at an early hour Sunday morning after the storm, and the water had completely receded, and it is dead certain that no one took a photograph before that time. The report that forty men perished in the power-house was exploded long before dawn on Sunday, and was generally known to be untrue before noon of that day. No man perished in the building.

The work of the New York Journal Relief Corps was greatly appreciated in Galveston, but the Journal's correspondents put some fancy touches to their reports concerning it. The Journal of September 15 had a page picture of "Mrs. Winnifred Black establishing a home and hospital for the bereaved little ones of the wrecked city." The Journal did establish a dispensary and hospital, but if Mrs. Black established such an institution for orphans the News reporters have not been swift enough to keep her pace. In

Sunday's Journal another picture of the same kind appears, and with it is a story in which Mrs. Black speaks of having spent the night in the Journal's dispensary and hospital. "The gas was out in the great theater," she writes, "and a few candles shed a flickering light as a nurse or doctor tiptoed down the ward." Galveston is one of those little cities which boasted of "the theater," but the theater went down in the storm. If Mrs. Black spent the night in the Journal's hospital she was sheltered by the partially-wrecked vestibule of the Ball High School building.

The correspondent of a Chicago paper sent a dispatch that cholera had broken out in Galveston. This, of course, was false. As stated by Dr. Donaldson, the chief surgeon of the New York Journal Relief Corps, there is surprisingly little sickness here.

General Scurry ordered two of the correspondents to leave town because of the untruthfulness of their reports, and this has had a very salutary effect.

As has previously been stated, there was plenty of wreck and ruin and horror, plenty of courage and heroism to be seen here to have taxed the pens of the most gifted writers to portray. There was no need for any correspondent to have gone outside the truth to secure material for a thousand thrilling stories, and there was no reason why any of them should have imposed upon the papers which employed them and sent them here to get the facts.

CHAPTER IV.

AFTER THE STORM.

GALVESTON THE SCENE OF ONE OF THE GREATEST CATASTROPHES IN THE WORLD'S HISTORY.

PEOPLE STAND DAZED AND HELPLESS.

A MEETING OF SURVIVORS IN TREMONT HOTEL BEGINS TO COPE WITH THE MIGHTY PROBLEM OF CLEARING AWAY THE DEBRIS AND REMOVING THE DEAD.

ONE IN EVERY FORTY A CORPSE.

FEEBLE ATTEMPT TO IDENTIFY BODIES AND TO BURY THEM—THE TASK TOO GREAT AND THEY ARE SENT TO SEA IN BARGES—FATE OF THE LOOTERS.

The horrors of Sunday were nothing as compared with those of Monday. Galveston had been the scene of one of the greatest catastrophes in the world's history.

There were stories of wonderful rescues and escapes, each of which at another time would be a marvel to the rest of the world, but in a time like this, when a storm so intense in its fury, so prolonged in its work of destruction, so wide in its scope and so infinitely terrible in its consequence has swept an entire city and neighboring towns for miles on either side, the mind cannot comprehend all the horror, cannot learn or know all of the dreadful particulars. One stands speechless and powerless to relate even that which he has felt and knows. Gifted writers have told of storms at sea, wrecking of vessels where hundreds were at stake and lost. That task pales to insignificance when compared with the task of telling of a storm which threatened the lives of perhaps 60,000 people, sent to their death perhaps 6000 people, and left others wounded, homeless and destitute and still others to cope with grave responsibility, to relieve the stricken, to grapple with and prevent the anarchist's reign, to clear the water-sodden land of putrefying bodies and dead carcasses, to perform tasks that try men's souls and sicken their hearts.

The storm at sea is terrible, but there are no such dreadful consequences as those which have followed the storm on the seacoast, and it is men who passed through the terrors of the storm, who faced death for hours, men ruined in property and bereft of families, who took up the herculean and well-nigh impossible task of bringing order out of chaos, of caring for the living and disposing of the dead before they made life impossible here. The storm came not without warning, but the danger which threatened was not realized, not even when the storm was upon the city. Friday night the sea was angry. Saturday morning it had grown in fury, and the wrecking of the beach resorts began. The waters of the Gulf hurried inland. The wind came at terrific rate from the north. Still men went to their business and about their work, while hundreds went to the beach to witness the grand spectacle which the raging sea presented. As the hours rolled on the wind gained in velocity and the waters crept higher and higher. The wind changed from the north to the northeast and the water came in from the bay, filling the streets and running like a millrace. Still the great danger was not realized. Men attempted to reach homes in carriages, wagon, boats or any way possible. Others went out in the storm for a lark. As the time wore on the water increased in depth and the wind tore more madly over the island. Men who had delayed starting for home, hoping for an abatement of the storm, concluded that the storm had grown worse and went out into that howling, raging, furious storm, wading through water almost to their necks, dodging flying missiles swept by a wind blowing 100 miles an hour. Still the wind increased in velocity, when, after it seemed impossible that it should be more swift, it changed from west to southeast, veering constantly, calming for a second and then coming with awful terrific jerks, so terrible in their power that no building could withstand them, and none wholly escaped injury.

Others were picked up at sea. And all during the terrible storm acts of the greatest heroism were performed. Hundreds and hundreds of brave men, as brave as the world ever knew, buffeted with the waves and rescued hundreds of their fellow-men. Hundreds of them went to their death, the death that they knew they must inevitably meet in their efforts. Hundreds of them perished after saving others.

NEAR 16TH AND M.

Men were exemplifying that supreme degree of love of which the Master spoke, "Greater love hath no man than this, that he give his life for his friend." Many of them who lost their lives in this storm in efforts to save their families, many to save friends, many more to help people of whom they had never heard. They simply knew that human beings were in danger and they counted their own lives as nothing.

The maximum velocity of the wind will never be known. The gauge at the Weather Bureau registered 100 miles an hour and blew away at 5.10 o'clock, but the storm at that hour was as nothing when compared with what followed, and the maximum velocity must have been as great as 120 miles an hour. The most intense and anxious time was between 8.30 and 9 o'clock, with raging seas rolling around them, with a wind so terrific that none could hope to escape its fury, with roofs beginning to roll away and buildings crashing all around them, men, women and children were huddled in buildings, caught like rats, expecting to be crushed to death or drowned in the sea, yet cut off from escape. Buildings were torn down, burying their hundreds, and were swept inland, piling up great heaps of wreckage. Hundreds of people were thrown into the water in the height of the storm, some to meet instant death, others to struggle for a time in vain, and thousands of others to escape death in most miraculous and marvelous ways.

Hundreds of the dead were washed across the island and the bay many miles inland. Hundreds of bodies were buried in the wreckage. Many who escaped were in the water for hours, clinging to driftwood, and landed bruised and battered and torn on the mainland.

The work of destruction began on the fateful Saturday of September 8, 1900. From 3 o'clock in the afternoon until 9 o'clock the same night, a period of six hours, the principal work of destruction was completed. In six brief hours the accumulations of many a lifetime had been swept away; thousands of lives had gone out, and the dismal Sabbath morning found a stricken population, paralyzed and helpless. In one of the silt-laden rooms of the Tremont Hotel a few citizens, headed by the mayor of the city, met to formulate some plans of relief. In a few hours the debris on Tremont street was cleared for a passage way from bay to Gulf, so that succor might reach the afflicted of the city. Those present at that meeting had not the faintest conception of what had occurred. They had gone through the storm, but beyond the knowledge that a number of lives had been sacrificed, with a serious destruction to property, they had not the remotest idea of the terrible calamity that had befallen the population. The burial of the dead was the first matter to engage the attention of that small meeting, and, as showing the lack of appreciation of the terrible situation, the suggestion was even discussed that the interment of the dead could not be legally performed without the assistance of a coroner's inquest! Temporary morgues were established for the identification of the dead, but so rapidly did these morgues fill up that even identification was out of the question. The whole population left were dazed beyond comprehension, and as the dreadful effects of the frightful hurricane were hour by hour becoming more fatally realized the exertions of the people rose with the occasion. At a subsequent meeting on the afternoon of the same sad day, called by word of mouth, committees were framed and the work of rescue and deliverance undertaken. What a task for a population cut off from all sources of communication with the outside world! In face of all these frightful conditions the spirit and resouces of the people never faltered or failed, and when the week is looked back upon it seems marvelous that human energy, stricken as it was with private and personal grief, could have accomplished what has been accomplished. To enumerate specially the work of individuals or organizations would be an impossibility at this time. The first and most efficient aid rendered by organized labor was the action of the Screwmen's Association, who tendered their services to the authorities for gratuitous, laborious work, accompanying the same with a private donation of $1000 for the relief of others. Think of that! To the credit of the colored screwmen's organization let it be said that they followed the splendid example of their white brethren with gratuitous labor, and, with asistance from all quarters, what remained of the city was soon placed in passable condition. There are heroes on the battlefield and on the mariner's deck, but no greater example of heroic effort was ever demonstrated by land or sea than has been exhibited by the population of Galveston during the past week. With sore hearts and desolated homes the population of the city rose to the emergency, and with a spirit that disaster cannot overcome, Galveston will go forward to a grander and better position than ever before. It has not been the courage of despair that has been evidenced, but the courage of determination to stay with Galveston, to make Galveston, and never to leave off until she takes the place that nature designed—one of the leading seaports of the American continent.

Sunday after the great hurricane it took the people of Galveston a whole day to realize what had happened. In fact, Monday was far spent before it dawned upon many citizens how widespread and awful the catastrophe really was. On Monday the question that passed from mouth to mouth was invariably: "Was there ever anything like it?" Appeals were made through all channels to the charitable world, and then Galveston went to work to bury her dead, to care for the injured and to disinfect the town. Three days passed in which the condition was almost chaotic, but on the fourth day order began to appear. Saturady night men stand amazed and now ask the question in a different vein, "Was there ever anything like it?" It is doubtful whether ever catastrophe was more complete over so great an area. It is doubtful whether ever a community arose so soon to the emergencies of the situation as has Galveston.

In the lower part of the city was a space four blocks wide and three miles long entirely denuded of houses. The wreckage and debris was washed up into the higher portions of the town, and even across the bay to Texas City and beyond. Among it were the bodies of hundreds of the dead who had lost their lives in that awful night of agony and fear. The first problem was the removal of these bodies, and the enormity of the task no one could foresee on that gloomy Monday morning.

There was no artificial light, for the gas plant was partially ruined. The water supply, brought in pipes sixteen miles from the mainland, had been interrupted. Houses, such as remained, were soaked with salt water. Wreckage was everywhere.

One person in every forty of the population was dead. Perhaps one in every ten was still suffering from cuts or bruises or shock received in the battle for life. The dead must be buried, the wounded and sick must be nursed, survivors must be housed, many of them in tents, until houses could be secured. They must be fed, and that twelve feet of salt water and downpour of rain upon unroofed houses had destroyed food in large quantities.

An attempt was made to bury the dead, but the ground was full of water, and it was impossible to dig trenches. Aldermen McMaster and M. P. Morrissey obtained authority to have the bodies taken to sea for burial, and a barge was brought up to the Twelfth street wharf for that purpose. The firemen rendered heroic service in bringing the bodies to the wharf, but it was almost impossible to get men to handle them.

Men stood about dazed and helpless. Few women and children were to be seen. Men saved themselves by struggling through the rushing water to places of safety. Women and children gave up after being buffeted in the waves and were drowned. A railroad man who went over a mile of track on the mainland saw, one after another, the bodies of nine girls and women. He covered them with bunches of sedge grass and clods to keep off the vultures until a burial party could be notified. There never will be a complete list of the dead. Even the Citizens' Committee did not know how many had been buried.

There was not time or means for burial on land. Parties acting independently went along the Gulf and bay beach for miles collecting the dead and burying them where they were found. On the mainland still other parties, some of them volunteers from Houston, performed the last sad offices for those who were swept out of their homes and across the bay. Until reports from all these could be had anything like the total number of bodies found could not be obtainable.

After that the lost and not found had to be dealt with to arrive at the mortality. From such indefinite returns as were made the conclusion was drawn that four-fifths of those who lost their lives were women and children. Loss of property was hardly to be mentioned in such close association with the mortality list. Galveston citizens were too busy with the burial of the dead and with the succor of the living to count the dollars that were gone.

But here was one of the best-built and most beautiful cities of its class. It had parks and hospitals and asylums and school buildings of the best architecture. It had great warehouses and docks and elevators. By some who looked about it was said that scarcely a structure in the city had escaped damage, and fully one-third were demolished. Representatives of the railroads which entered Galveston hastened to the scene of destruction. The conclusion was that hardly within three weeks would they be able to get a train into the city.

Every hour the situation changed for the worse, and the mind became dazed midst the gruesome scenes—dead, dead, dead, dead everywhere. The bodies of human beings, the carcasses of animals, were strewn on every hand. There were countless numbers of bodies. The bay was filled with them. Like jelly-fish, the corpses were swept with the changing tide. Here a face protruded above the water; there the stiffened hand of a child; here the long, silken tresses of a young girl; there a tiny foot, and just beneath the glassy surface of the water full outlines of swollen bodies were visible.

Such scenes drove men and women to desperation and insanity. A number sought freedom in the death which they fought so stoutly. A young girl, who survived to find mother, father and sisters dead, crept far out on the wreckage and threw herself into the bay. Only a few viewed this act of desperation. It was of no consequence. What was one death more or less? What would the hundreds or less on the ill-fated island care?

During the storm and afterward a great deal of looting was done. Many stores had been closed, their owners leaving to look after their families. The wind forced in the windows and left the goods prey for the marauders. Ghouls stripped the dead bodies of jewelry and articles of value. Captain Rafferty, commanding United States troops here, was applied to for help, and he sent in seventy men, the remnant of the battery of artillery, to do police duty. Three regiments were sent from Houston and the city placed under martial law. Hundreds of desperate men roamed the streets, crazed with liquor, which many had drank because nothing could be obtained with which to quench their thirst. Numberless bottles and boxes of intoxicating beverages were scattered about and easy to obtain.

Robbing and rioting were going on during the night, and as the town was in darkness, the efforts of the militia to control the lawless element were not entirely successful. Big bonfires were built at various places from heaps of rubbish to enable troops the better to see where watchfulness was needed. Reports said that from fifty to 100 looters and vandals were slain in the city and along the island beach.

The most rigid enforcement of martial law was not able to suppress it entirely. Adjutant-General Scurry's men arrested 100 or more looters. Thirty-three negroes, with effects taken from dead bodies, were ordered to be tried by court-martial. They were tried, convicted and ordered to be shot. One negro had twenty-three human fingers with rings on them in his pocket.

Decomposition was rapidly setting in, and the important question was the one of the disposition of the bodies. In great wagon-loads some were hauled from the streets and placed on barges, which were taken out beyond the jetties

FROM PIER 23, LOOKING EAST.

to the deep waters of the Gulf. Others, too far decomposed for human hands to touch, were left in heaps, with inflammable rubbish piled liberally upon and about them, and the torch did the rest.

An eye-witness of that awful morning said: "I was going to take the train at midnight, and was at the depot when the worst of the storm came up. There were 150 people in the depot. We all sat there for nine hours. The back part of the depot blew in Sunday morning. I went back to the Tremont Hotel at 9 o'clock, and the streets were literally filled with dead and dying people. All morning they had no rations except beer, which was very warm, and oyster crackers.

"The Sisters' Orphan Hospital was a terrible scene. I saw there over ninety dead children and eleven dead Sisters. We took the steamer Allen Charlotte across the bay up Buffalo bay over to Houston in the morning, and I saw fully fifty dead swollen bodies floating in the water. I saw and counted in all over 800 dead. The stench in some of the streets was so terrible I could not pass along them without it making me deathly sick. I saw one dray with sixty-four dead bodies being drawn by four horses to the wharves, where the bodies were unloaded on a tug and taken out in the Gulf for burial."

An awful epidemic was feared. The stench became so great that hundreds died from sickness and neglect. For a time there was menace of a water famine, but one of the water mains was found and tapped, which furnished a supply.

It was found impossible to load all the dead on tugs for burial at sea, and it became necessary to place them in heaps, with debris to cover them, and to burn them. After Tuesday the majority of bodies found were thus cremated.

CHAPTER V.

SURVEY OF THE WRECKAGE.

PATH OF THE STORM'S FURY THROUGH GALVESTON CLEARLY MARKED BY A RIDGE OF DEBRIS.

SIXTY-SEVEN BLOCKS DEVASTATED.

IN ONE SECTION OF THE CITY THIS AREA WAS SWEPT CLEAN, SAVE FOR THE RUBBISH THAT IS PILED UP IN MASSIVE HEAPS ALL OVER IT.

BEACH AS BARE AS IT WAS A CENTURY AGO.

BATHHOUSES, THE PAGODAS, CONCERT HALLS AND THE MIDWAY WERE SWEPT INTO THE GULF, AND THE WATER NOW ROLLS OVER THE SAND WHERE THEY STOOD.

Some idea of the extent of the destructive path of the hurricane can be got from a view of the beach front east of Tremont street. Standing on the high ridge of debris that marks the line of devastation extending from the extreme west end to Tremont street an unobstructed view of the awful wreckage is presented.

Drawing a line on the map of the city from the center of Tremont street and Avenue P straight to Broadway and

Thirteenth street, where stands the partly demolished Sacred Heart Church, all the territory south and east of this line is leveled to the ground.

The ridge of wreckage of the several hundred buildings that graced this section before the storm marks this line as accurately as if staked out by a surveying instrument. Every building within the large area was razed by the wind or force of the raging waters, or both.

This territory embraces sixty-seven blocks, and was a thickly-populated district. Not a house withstood the storm, and those that might have held together if dependent upon their own construction and foundations were buried beneath the stream of buildings and wreckage that swept like a wild sea from the east to the west, demolishing hundreds of homes and carrying the unfortunate inmates to their death, either by drowning or from blows of the flying timbers and wreckage that filled the air. The strongest wind blew later in the evening, when it shifted to the southeast and attained a velocity of from 100 to 120 miles an hour. The exact velocity was not recorded, owing to the destruction of the wind gauge of the United States Weather Bureau after it had registered a 100-miles-an-hour blow for two minutes. This terrific southeast wind blew the sea of debris inland and piled it up in a hill ranging from ten to twenty feet high, and marking the northern line of the storm's path along the southeastern edge of the island.

In one place near Tremont street and Avenue P four roofs and remnants of four houses are jammed within a space of about twenty-five feet square. Beneath this long ridge many hundred men, women and children were buried, and cattle, horses and dogs and other animals were piled together in one confused mass. While every house in the city and suburbs suffered more or less from the hurricane and encroachment of the Gulf waters, the above section suffered the most in being swept as clean as a desert. Another area extending east to Thirteenth street and south of Broadway to the Gulf suffered greatly, and few of the buildings withstood the storm, none without being damaged to a more or less extent. From Tremont street and Avenue P$\frac{1}{2}$ the wind came northward for about two blocks and then cut across westward to the extreme western limits of the city; in fact, swept clear on down the island for many miles. The path of the leveled ground west from Avenue P cleared the several blocks extending south to the beach and west to Twenty-seventh street. It cut diagonally southwest on a straight line within three blocks of the beach and down west on the beach many miles beyond the city limits. This does not mean that the path of the storm was confined to this stretch of territory—not by any means. There were many blocks in the center of the city almost totally demolished by the fury of the wind and sea, but the above long line of about four miles of the city proper and many miles of country lands were swept clean of buildings and all other obstructions.

A few of the piles that once supported the railway trestle extending from Center street to Tremont street on the beach are all that remains to mark the curved line of right of way.

Not a vestige of the three large bathhouses of O'Keefe's Pagoda and Murdock is to be seen.

The midway, with its many old shacks and frame houses, concert halls and other resorts, was swept to the sea, and the Gulf now plays twenty feet north of where the midway marked the beach line. The Olympia-by-the-Sea likewise fell an early prey to the storm, and the surf which formerly kissed the elevated floor of the Olympia now sweeps across the electric railway track about fifteen feet north of the big circular building. On Tremont street and Avenue P½ two buildings stand, or rather two structures mark where two frame buildings battled with the raging elements. Both houses were stripped of every piece of furniture, wallpaper, window-frames and doors on the first floor and the second floors remained intact. The houses were blown down from their elevated foundations and dropped down on the ground, and the sea washed the interior of the first floors almost up to the ceilings. The families took refuge in a house across the street, which gave way and was leveled almost to the ground, but all the inmates escaped with their lives. These two dwellings stand like charmed structures in the center of the hurricane's track.

Galveston's citizens rallied heroically to the emergency. They organized into a great camp, practically under martial law. Uncle Sam's regulars, those that survived the destruction of the barracks; the police force and the city military patrol the streets. Companies were formed to take up every branch of the work of relief and restoration. None of the help which came from outside was wasted or went amiss, but Galveston was facing some problems that morning which even a nation's charity could not reach. The dead must be disposed of, and that quickly in the hot climate. On the morning after the storm subsided the work of collecting bodies began.

It was also found that duty to the living demanded immediate action. Many corpses were without clothing sufficient to permit identification. When the attempt was made to dig graves in some places it was discovered that the water-soaked condition of the ground made burial in mother earth impossible, so the sea, which had wrought such havoc, received its victims. Johnstown was horrible when the first shock was over and the population of the charnel-houses began to grow, but to Johnstown flocked volunteers by hundreds and men accustomed by vocation to handle the dead in gruesome form. Galveston, with every bridge gone, with only an occasional tug or sloop to give communication with the mainland, was forced to meet alone this first emergency of caring for the dead and protecting the living. It was work that called for stout hearts. Many shrank from it. Bodies were gathered in two-wheeled carts and hauled to a barge, which was turned into a morgue. So far as possible a list was kept, and it soon grew to appalling proportions.

As was the case at Johnstown, the number of the dead at Galveston will never be known accurately. When the waters came they reached a height of twelve feet above the ordinary tide. That meant three to five feet depth on the more elevated parts of the island. Sweeping and whirling through the streets, the waves carried the weak and panic-stricken with them, and when the water receded rapidly in the early hours of Sunday morning they bore the dead out into the Gulf on one side and into the bay on the other. Many of those thus taken away were never found. Some were picked up along the mainland and were given hasty interment. Workers in the immense piles of rubbish got down to systematic operation under the direction of the State troops, commanded by Adjutant-General Scurry. An eye-witness relates what he saw in these words:

"The workers are just going in among the dead at Galveston, and the corpses are being piled up in big rows, like cord wood. There are no coffins or undertakers' supplies, and great loads of dead are hastened off and dumped into the sea. There is no chance to bury them on the island. Rich men, poor men, white men and black men are all mixed up and all treated the same way. All corpses look alike to these workers in the vast heaps of dead."

A terse telegram from Dallas, the news of which was carried thither by some refugee of the fated city, tells vividly the conditions which prevailed the second morning after the catastrophe:

"The situation grows worse every minute; water and ice are needed. People in frenzy from suffering from these causes. Scores have died since last night, and a number of sufferers have gone insane."

The true nature of the appalling tragedy grew steadily

NEAR L STREET.

upon the minds, first of the survivors at Galveston, and then upon the outside world, as further inroads were made into that endless pile of debris and as communication slowly opened up over the broken railroads and prostrate telegraph lines. Each refugee reaching the territory beyond the path of the storm brought with him stories that almost surpassed belief, and afforded proof of a calamity unparalleled in American history.

When early reports of the disaster reached the outside world, giving an estimate of 1500 souls swept suddenly into eternity, the news was received with incredulity. Monday night Mayor Jones of Galveston made the astounding statement that a conservative estimate of the dead was not less than 5000 souls. The earliest news from the stricken city was as sensational as it was shocking. So dangerous a menace to the safety of the survivors had the existence of rapidly decomposing corpses become that every able-bodied man who could lend a hand was engaged in the work of cremating the bodies that remained in the debris and of consigning to the sea or the common trenches those that were picked up on the streets or along the storm-swept beach. Already it was stated 2300 were thus disposed of in the interest of human life and the preservation of public sanitation.

Though hardly credible, the news was amply confirmed that fiends in human form flocked to Galveston to rob the dead and to loot their homes. Intense indignation was aroused over the terrible disclosures, and the report of the killing of more than a score of the ghouls by the soldiers and citizens found the strongest commendation in public sentiment. Mayor Jones took a firm stand on the situation, and had both the support of State and regular troops. It was assured that the city could be more thoroughly policed and further desecration of the dead promptly stopped.

The most exhaustive efforts were made to obtain a complete list of the victims of the disaster. That was an impossible task, and only through a new census of the living could the full number and the identity of those who lost their lives be known.

Notwithstanding the summary punishment meted out to looters and ghastly robbers of the dead, ghouls continued to hold their orgies. The majority of these men were negroes, but there were also whites who took part in the desecration. Some of them were natives, and some had been allowed to go over from the mainland under the guise of "relief" work. Not only did they rob the dead, but they mutilated bodies in order to secure their booty. A party of ten negroes were returning from a looting expedition. They had stripped corpses of all valuables, and the pockets of some of the looters were fairly bulging out with fingers of the dead which had been cut off because they were so swollen the rings could not be removed.

During the robbing of the dead not only were fingers cut off, but ears were stripped from the head in order to secure jewels of value. A few government troops at first assisted in patrolling the city. Private citizens also endeavored to prevent the robbing of the dead, and on several occasions killed the offenders.

The city of Galveston strained every nerve to clear the ground and secure from beneath the debris the bodies of human beings and animals and to get rid of them. It was a task of great magnitude and was attended with untold difficulties. There was a shortage of horses to haul the dead, and there was a shortage of willing hands to perform the gruesome work. Tuesday noon it became apparent that it would be impossible to bury the dead even in trenches, and arrangements were made to take them to sea. Barges and tugs were quickly made ready for the purpose, but it was difficult to get men to do the work. The city's firemen worked hard in bringing bodies to the wharf, but outside of them there were few who helped. Soldiers and policemen were accordingly sent out, and every able-bodied man they found was marched to the wharf front. The men were worked in relays, and were supplied with stimulants to nerve them for the task. At nightfall three bargeloads, containing about 700 human bodies, were sent to sea, where they were sunk with weights. Darkness compelled suspension of the work until morning. Toward night great difficulty was experienced in handling the bodies of negroes, which were badly decomposed. No effort was made after 9 o'clock Tuesday morning to place the bodies in morgues for identification, for it was imperative that the dead should be gotten to sea as soon as possible. Many of the bodies taken out were unidentified. They were placed on the barges as quickly as possible, and lists were made while the barges were being towed to sea.

A large number of dead animals were hauled to the bay and dumped in, to be carried to sea by the tides. One hundred and twenty-five men worked all day Tuesday and Tuesday night in uncovering the machinery of the water-works from the debris. It was hoped that it would be possible to turn on the water for a while, and it was planned to set fire to the debris and cremate the bodies buried under it.

Mayor Jones gave very full scope to Chief of Police Ketchum and J. H. Hawley, chairman of the Committee on Public Safety, to swear in citizens as officers.. Picket lines were established around the large stores and guards placed on duty. Soldiers and police were instructed to shoot anyone caught looting or attempting to loot. As the work of collecting the bodies proceeded it became apparent that the death list would run much higher than was at first supposed. Conservative estimates on Tuesday placed the number of dead in the city at 5000. The mainland, Galveston Island and Bolivar peninsula were bestrewn with dead.

A relief train from Houston with 250 men on board and two carloads of provisions came down over the Galveston, Houston & Northern Railroad Monday to a point about five miles from Virginia Point. It was impossible for them to get the provisions or any considerable number of the men to Galveston, so they turned their attention to burying the dead lying around the mainland country. There was no fresh-water famine after Monday, as the pipes from the supply wells were running at the receiving tanks. It was difficult, however, to get it to parts of the city where it was needed.

Col. L. J. Polk, general manager of the Gulf, Colorado & Santa Fe Railroad, stated that all of the bridges across Galveston bay were gone, nothing remaining but the piles. He said it would take ten days or two weeks to restore rail communication to Galveston, provided work could be instituted at once from the mainland.

Refugees crossing over the bay and by way of Texas City Junction reaching Houston told harrowing stories that varied only in detail. Instead of dissipating the gloom which had settled over all since Monday, they had nothing to say except what made the situation appear worse as time progressed. What was most needed was means of transportation across the bay to carry over the homeless and the destitute. With regard to the resumption of direct rail communication with the island, no one could tell when it would take place. The Santa Fe, the Missouri, Kansas & Texas, the International & Great Northern and the Galveston, Houston & Henderson decided to construct a temporary bridge for joint use, and the Southern Pacific joined eventually in the enterprise.

Regarding the movement of export freight, no definite plans could be agreed upon. The all-rail routes to the ports of Sabine Pass and New Orleans had to be used. Detachments of militia were stationed at Texas City and Virginia Point to prevent the passage of persons who had no business in Galveston and whose presence would prove an incumbrance at that time. Eighty-three bodies were buried at Texas City. All of these floated in from the island. Two hundred bodies were recovered and buried at Virginia Point.

The following statement of conditions at Galveston and appeal for aid was issued by the Local Relief Committee:

"A conservative estimate of the loss of life is that it will reach 5000; at least 5000 families are shelterless and wholly destitute. The entire remainder of the population is suffering in greater or less degree. Not a single church, school or charitable institution, of which Galveston had so many, is left intact. Not a building escaped damage, and half the whole number were entirely obliterated. There is immediate need for food, clothing and household goods of all kinds. If nearby cities will open asylums for women and children the situation will be greatly relieved. Coast cities should send us water as well as provisions, including kerosene oil, gasoline and candles. W. C. Jones, mayor; M. Lasker, president Island City Savings Bank; J. D. Skinner, president Cotton Exchange; C. H. McMaster, for Chamber of Commerce; R. G. Lowe, manager Galveston News; Clarence Owsley, manager Galveston Tribune."

Estimates of the number of dead varied at this time, but Mayor Jones of Galveston declared that it was 5000, and even the most intelligent survivors of the tragedy could not guess what the property damage was. Some said $20,000,000, but that was only a number given to express a vague idea of material losses, which were countless and could never be recovered.

At the urgent call of citizens of Galveston three regiments of soldiers came. The troops were placed on duty patrolling the streets, and looting gradually ceased and the people of the stricken city began to recover from the first effects of the disaster and to face the future and the enormous duties at hand in clearing away the wreckage and gathering up the dead. The stench from the bodies already began to fill the air.

Mr. Wortham, ex-secretary of state, who accompanied General Scurry in command of the militia, returned to Austin after a thorough examination of the city, and made this statement:

"The situation at Galveston beggars description. I am convinced that the city is practically wrecked for all time to come.

"Fully 75 per cent. of the business portion of the town is irreparably wrecked, and the same percentage of damage is to be found in the residence district. Along the wharf front great ocean steamers have bodily bumped themselves on the big piers and lie there, great masses of iron and wood that even fire cannot totally destroy. The great warehouses along the water front are smashed in on one side, unroofed and gutted throughout their length, their contents either piled in heaps on the wharves or along the streets. Small tugs and sailboats have jammed themselves into buildings, where they were landed by the incoming waves and left by the receding waters.

"Houses are packed and jammed in great confusing masses in all of the streets. Great piles of human bodies, dead animals, rotting vegetation, household furniture and fragments of the houses themselves are piled in confused heaps right in the main streets of the city. Along the Gulf front human bodies are floating around like cordwood. Intermingled with them are to be found the carcasses of horses, chickens, dogs and rotting vegetable matter. Above all rises the foulest stench that ever emanated from any cesspool, absolutely sickening in its intensity and most dangerous to health in its effects.

"Along the Strand, adjacent to the Gulf front, where are located all the big wholesale warehouses and stores, the situation almost defies description. Great stores of fresh vegetables have been invaded by the incoming waters, and are now turned into garbage piles of most befouling odors. The Gulf waters while on the land played at will with everything, smashing in doors of stores, depositing bodies of human beings and animals where they pleased, and then receded, leaving the wreckage to tell its own tale of how the work had been done. As a result the great warehouses are tombs wherein are to be found dead bodies of human beings and carcasses almost defying the efforts of relief parties.

"In the piles of debris along the street, in the water and scattered throughout the residence portion of the city are masses of wreckage, and in these great piles are to be found

GALVESTON WHOLESALE HAT AND SHOE COMPANY, FOUR-STORY BRICK.

BURNING THE DEAD.

more human bodies and household furniture of every description."

Beginning on Monday morning after the storm, some effort was made to place the dead in rows, so that they could be identified and buried. But this lasted less than a day. In two days no effort was made toward identification, and the citizens, soldiers, firemen and every man capable of work were pressed into the service of placing the bodies on boats or of throwing them into heaps of burning timber, where they were cremated. On Friday, nearly a week after the deluge, the Houston Post printed the names of 2701 persons whom it was known were lost.

In one of the Catholic parishes 600 had perished. As time passed on the terrible truth was pressed home on the minds of the people that the mortality by the storm had possibly reached 8000, or nearly one-fourth of the population. The exact number will never be known, and no list of the dead will be accurately made out, for the terrible waters carried to sea and washed on distant and lonely shores many of the bodies. The unknown dead of the Galveston horror will probably forever surpass the number of those who are known to have perished in that awful night, when the tempest raged and the storm was on the sea, piling the waters to unprecedented heights on Galveston Island.

Men worked for many days clearing away the debris, and the task was slowly accomplished. Men worked at the bayonet point pulling and hauling at the piles of rubbish and **throwing it on the fires, where the decomposed bodies of the** dead were thrown as they were extricated from the ruins. At one time it was suggested that the vast mass of rubbish in certain places should be set on fire and the work of clearing the streets and vacant lots accomplished in that way, but it was thought too dangerous, lest the fiend of fire should finish what that of the flood had begun.

The editor of the Galveston News sounded the note of warning to his fellow-townsmen, and begged that care should be exercised in clearing away the wreckage, so that what it contained that was valuable might be saved. He said:

"Suggestions have been made to burn the piles of lumber of all kinds in the rafts, but this seems both impracticable and inadvisable. If it can be secured, every stick and board will be of use hereafter. The only reason for burning the rafts given are that it will cremate the bodies of the dead known to be in them, and some are supposed to be in almost all of them.

"Sickness resulting from the decaying bodies is predicted if this is not done. But if it is attempted more loss of life is likely to occur from it than will result from sickness arising from putrid bodies. Once let the fire demon get hold of the immense masses of lumber and the remaining portion of the city may be wiped out. No one who has seen a conflagration in a city can doubt that all the fire apparatuses in Texas would be ineffectual to stop the march of the flames to the bay in case of a strong south wind. Many houses partially wrecked are in the piles, and many household goods belonging to people who have lost all may be recovered. Disinfect the rafts as far as possible and remove the lumber. Preserve it as far as can be done conveniently. It will be needed for building temporary homes for the destitute.

"We have thousands of homeless people in the city, and while free transportation is offered to those who wish to go, there are many who have no friends to go to. These people must be cared for. Some are now crowded in the homes of friends and others are located in the large buildings in the business district. All are only temporarily provided for. Something must be done to house them at least temporarily when cold weather approaches. It would be well to issue permits for temporary buildings to be erected from the debris of wrecked homes without regard to the fire rules of the city as they now stand, but with the distinct proviso that they should be removed on a certain date. I am no advocate of ramshackle shanties as permanent buildings in the city— in any part of it. But I appreciate the fact that we are facing an emergency that requires prompt action to prevent severe suffering in the near future. Galveston people have not in the past turned their faces against the suffering poor, and I do not think they will do so in the future.

"Substantial buildings should be required in permanent structures. There is no reason why the wreckage should not be used in erecting temporary shelter for the homeless. Lumber promises to be a scarce article when once the resumption of building is begun, and every board, rafter and scantling in the pile of wreckage should be saved.

"There is valuable wreckage through the rafts. There are desks and trunks that may contain papers of value to the owners, but valueless to others. These should be placed aside and saved for identification by their owners. Articles of personal apparel may sometimes be the means of settling the estates of the dead. Many wills may be found stored away in frail desks that by some chance may have escaped total wreckage in the storm. Jewelry and personal ornaments are not unlikely to be found in places where least expected. People fleeing from wrecked houses do not stop to search in trunks for jewel boxes. Many of these doubtless remain in the mass of chaos-like wreckage and may be recovered as the piles are cleared away."

CHAPTER VI.

STORY AS TOLD BY THE FIRST NEWSPAPER WOMAN TO VISIT THE RUINS OF THE CITY.

I begged, cajoled and cried my way through the line of soldiers with drawn swords who guard the wharf at Texas City and sailed across the bay on a little boat which is making irregular trips to meet the relief trains from Houston.

The engineer who brought our train down from Houston spent the night before groping around in the wrecks on the beach looking for his wife and three children. He found them, dug a rude grave in the sand and set up a little board marked with his name. Then he went to the railroad company and begged them to let him go to work.

The man in front of me on the car had floated all Monday night with his wife and mother on a part of the roof of his little home. He told me that he kissed his wife good-by at midnight and told her could not hold on any longer; but he did hold on, dazed and half conscious, until the day broke and showed him that he was alone on his piece of dried wood. He did not even know when the women that he loved had died.

Every man on the train—there were no women there—had lost some one that he loved in the terrible disaster, and was going across the bay to try and find some trace of his family—all except the four men in my party. They were from outside cities—St. Louis, New Orleans and Kansas City. They had lost a large amount of property and were coming down to see if anything could be saved from the wreck.

They had been sworn in as deputy sheriffs in order to get into Galveston. The city is under martial law, and no human being who cannot account for himself to the complete satisfaction of the officers in charge can hope to get through.

'TWAS HIS WIFE AND CHILDREN BURNING.

We sat on the deck of the little steamer. The four men from out-of-town cities and I listened to the little boat's wheel plowing its way through the calm waters of the bay. The stars shone down like a benediction, but along the line of the shore there rose a great leaping column of blood-red flame.

"What a terrible fire!" I said. "Some of the large buildings must be burning." A man who was passing the deck behind my chair heard me. He stopped, put his hand on the bulwark and turned down and looked into my face, his face like the face of a dead man, but he laughed.

"Buildings?" he said. "Don't you know what is burning over there? It is my wife and children, such little children; why, the tallest is not as high as this"—he laid his hand on the bulwark—"and the little one was just learning to talk.

"She called my name the other day, and now they are burning over there, they and the mother who bore them.

She was such a little, tender, delicate thing, always so easily frightened, and now she's out there all alone with the two babies, and they're burning them. If you're looking for sensations there's plenty of them to be found over there where that smoke is drifting."

The man laughed again and began again to walk up and down the deck.

"That's right," said the United States marshal of Southern Texas, taking off his broad hat and letting the starlight shine on his strong face. "That's right. We had to do it. We've burned over 1000 people today, and tomorrow we shall burn as many more.

"Yesterday we stopped burying the bodies at sea. We had to give the men on the barges whiskey to give them courage to do their work. They carried out hundreds of the dead at one time, men and women, negroes and white people, all piled up high as the barge could stand it, and the men did not go far enough out to sea, and the bodies have been drifting back again."

"Look!" said the man who was walking the deck, touching my shoulder with his shaking hand. "Look there!"

GHASTLY OBJECTS FLOATING IN THE WATER.

Before I had time to think I did look, and I saw floating in the water the body of an old, old woman, whose hair was shining in the starlight. A little further on we saw a group of strange driftwood. We looked closer and found it to be a mass of wooden slabs with names and dates cut upon them, and floating on top of them were marble stones, two of them.

The graveyard, which has held the sleeping citizens of Galveston for many, many years, was giving up its dead. We pulled up at a little wharf in the hush of the starlight. There were no lights anywhere in the city, except a few scattered lamps shining from a few desolate, half-destroyed houses. We picked our way up the street. The ground was slimy with the debris of the sea. Great pools of water stood in the middle of the street.

We climbed over wreckage and picked our way through heaps of rubbish. The terrible, sickening odor almost overcame us, and it was all that I could do to shut my teeth and get through the streets somehow.

The soldiers were camping on the wharf front, lying stretched out on the wet sand, the hideous, hideous sand, stained and streaked in the starlight with dark and cruel blotches. They challenged us, but the marshal took us through under his protection. At every street corner there was a guard, and every guard wore a six-shooter strapped around his waist.

NO AMERICANS AMONG THE GHOULS.

"The best men!" said the marshal. "They've left their own misery and come down here to do police duty. We needed them. They had to shoot twenty-five men yesterday for looting the dead. Not Americans, not one of them. I saw them all—negroes and the poor whites from Southern Europe. They cut off the hands of their victims. Every

IN FRONT OF CATHOLIC CHURCH ON BROADWAY.

citizen in Galveston has orders to shoot without notice any one found at such work."

We got to the hotel after some terrible nightmare-fashioned plodding through dim streets like the line of forlorn ghosts in a half-forgotten dream. At the hotel, a big, typical Southern hotel, with a dome and marble rotunda, the marble stained and patched with the sea slime, the clerk told us that he had no rooms. We tried to impress him in some way, but he would not look up from his book, and all he said was "No room" over and over again like a man talking in his sleep.

We hunted the housekeeper and found there was room and plenty of it, only the clerk was so dazed that he did not know what he was doing. There was room, but no bedding and no water and no linen of any sort.

General McKibben, commander in charge of the Texas division, was downstairs in the parlor reading dispatches, with an aide and an orderly or two at his elbow. He was horrified to see me.

"How in the world did you get here," he said. "I would not let any woman belonging to me come into this place of horror for all the money in America. I am an old soldier, madame. I have seen many battlefields, but let me tell you that since I rowed across the bay the other night and helped the man at the boat steer to keep away from the floating bodies of dead women and little children, I have not slept one single instant.

"I have been out on inspection all day, and I find that our

first estimate of the number of dead was very much under the real. Five thousand would never cover the number of people who died here in that terrible storm.

"I saw my men pulling away some rubbish this very morning right at the corner of the principal street. They thought there might be some one dead person there. They took out fourteen women and three little children. We have only just begun to get a faint idea of the hideous extent of this calamity. The little towns along the coast had been almost completely washed out. We hear from them every now and then as some poor, dazed wretch creeps somehow into shelter and tries to tell his pitiful story. We have only just begun our work.

MOST URGENT NEED FOR DISINFECTANTS.

"The people all over America are responding generously to our appeals for help, and I would like to impress it upon them that what we need now is money, money, money and disinfectants. Tell your people to send all the quicklime they can get through. I wish I could see a dozen trainloads of disinfectants landed in this city tomorrow morning. What we must fight now is infection, and we must fight it quick and with determination or it will conquer us."

The men of my party came over and took me from the great damp tomb of a room, where I was trying to write, to the Aziola Club across the street.

There were eighteen or twenty men there, most representative men of the city of Galveston, rich, influential citizens. They had all been on police duty or rescue work of some sort. The millionaire at the table next to me wore a pair of workmen's brogans, some kind of patched old trousers and a colored shirt much the worse for wear. He had been directing a gang of workmen who were extricating the dead from the fallen houses all day long.

The man on my right had lain for four hours under a mass of rubbish on Monday and had heard his friends pass by and recognized their voices, but could not groan loud enough for them to hear him. He told us what he was thinking about as he lay there with a man pinned across his chest and two dead men under him. He tried to make his story amusing and we all tried to laugh.

Every man in that room had lost nearly every dollar he had in the world, and two or three of them had lost the nearest and dearest friends they had on earth, but there were no sighs, and there was not one man who spoke in anything but tones of courageous endurance. In the short time I have been here I have met and talked with women who saw every one they loved on earth swept away from them out into the storm.

I have held in my arms a little lisping boy not eight years old, whose chubby face was set hard when he told me how he watched his mother die. But I have not seen a single tear. The people of Galveston are stunned with the merciful bewilderment which nature always sends at such a time of sorrow.

NO WIRE INTO GALVESTON.

As I look out of my window I can see the blood-red flame leaping with fantastic gesture against the sky. There is no wire into Galveston, and I will have to send this message out by the first boat. The Western Union hopes to get its wires through this afternoon. Then I will have the situation better in hand and will be able to tell more definitely just what this brave people, who are trying so courageously to stem the awful tide of misery which has overwhelmed them, need the most.

For the present the two things needed are money and disinfectants. More nurses and doctors are needed. Galveston wants help—quick, ready, willing help. Don't waste a minute to send it. If it does not come soon this whole region will be a prey to a plague such as has never been known in America. Quicklime and disinfectants and money and clothes—all these things Galveston must have, and have at once, or the people of this country will have a terrible crime on their conscience.

The people of Galveston are making a brave and gallant fight for life. The citizens have organized under efficient and willing management. Gangs of men are at work everywhere removing the wreckage. The city is districted according to wards, and in every ward there is a relief station. They give out food at the relief stations. Such food as they have will not last long.

I sat in one relief station for half an hour this morning and saw several people who had come asking for medicine and disinfectants and a few rags of clothing to cover their pitiful nakedness turned away. The man in charge of the bureau took the last nickel he had in the world out of his pocket and gave it to make up a sum for a woman with a newborn baby in her arms to buy a little garment to cover its shivering flesh.

The people of the State of Texas have risen to the occasion nobly. They have done everything that human beings, staggering and dazed under such a terrific blow, could possibly do, but they are only human. This is no ordinary catastrophe. No one who has not been here to see with his own eyes the awful havoc wrought by the storm can realize the tenth part of the misery these people are suffering.

NO SEARCH NEEDED TO FIND THE MISERY.

I asked a prominent member of the citizens' committee this morning where I should go to see the worst work which the storm had done. He smiled at me a little pitifully. His house, every dollar he has in the world and his children were swept away from him last Sunday night.

"Go?" said he. "Why, anywhere within two blocks of the very heart of the city you will see misery enough in half an hour to keep you awake for a week of sleepless nights."

I went toward the heart of the city. I did not know what the names of the streets were or where I was going. I simply picked my way through masses of slime and rubbish which scar the beautiful wide streets of this once beautiful city. They won't bear looking at, those piles of rubbish. There are things there that gripe the heart to see—a baby's shoe, for instance, a little red shoe, with a jaunty tasseled lace; a bit of a woman's dress and letters. Oh, yes, I saw these things myself, and the letter was wet and grimed with the marks of the cruel sea, but there were a few lines legible in it. "Oh, my dear," it read; "the time seems so long. When can we expect you back?" Whose hand had written or whose had received no one will ever know.

The stench from these piles of rubbish is almost overpowering. Down in the very heart of the city most of the dead bodies have been removed, but it will not do to walk far out. Today I came upon a group of people in a by-street, a man and two women, all colored. The man was big and muscular. One of the women was old and one was young. They were digging in a heap of rubbish, and when they heard my footsteps the man turned an evil, glowering face upon me and the young woman hid something in the folds of her dress. Human ghouls, these, prowling in search of prey.

A moment later there was noise and excitement in the little narrow street, and I looked back and saw the negro running, with a crowd at his heels. The crowd caught him and would have killed him, but a policeman came up. They tied his hands and took him through the streets with a whooping rabble at his heels. It goes hard with a man in Galveston caught looting the dead in these days.

SHOT FOR CUTTING DEAD WOMAN'S EARS.

A young man well known in the city shot and killed a negro who was cutting the ears from a dead woman's head to get her earrings out. The negro lay in the street like a dead dog, and not even the members of his own race would give him the tribute of a kindly look.

The abomination of desolation reigns on every side. The big houses are dismantled, their roofs gone, windows broken, and the high-water mark showing inconceivably high on the paint. The little houses are gone—either completely gone as if they had been made of cards and a giant hand which was tired of playing with them had swept them all off the board and put them away, or they are lying in heaps of kindling wood, covering no one knows what horrors beneath.

The main streets of the city are pitiful. Here and there a shop of some sort is left standing. South Fifth street looks like an old man's jaw, with one or two straggling teeth protruding. The merchants are taking their little stores of goods that have been left them and are spreading them out in the bright sunshine, trying to make some little husbanding of their small capital.

PEOPLE ARE HARDENED TO THE HORRORS.

They will stand and tell the most hideous stories, stories that would turn the blood in the veins of a human machine cold with horror, without the quiver of an eyelid. A man sat in the telegraph office and told me how he had lost two Jersey cows and some chickens. He went into minute particulars, told how his house was built and what it cost and how it was strengthened and made firm against the weather. He told me how the storm had come and swept it all away and how he had climbed over a mass of wabbling roofs and found a friend lying in the curve of a big roof, in the stoutest part of the tide, and how they two had grasped each other and what they said.

He told me just how much his cows cost and why he was so fond of them, and how hard he had tried to save them, but I said: "You have saved yourself and your family; you ought not to complain." The man stared at me with blank, unseeing eyes. "Why, I did not save my family," he said. "They were all drowned. I thought you knew that. I don't talk very much about it."

The hideous horror of the whole thing has benumbed every one who saw it. No one tells the same story of the way the storm arose or how it went. No man tells the story of his rescue quite alike. I have just heard of a little boy who was picked up floating on a plank. His mother and father and brothers and sisters were all lost in the storm. He tells a dozen different stories of his rescue on the night of the storm.

But the city is gradually getting back to a normal understanding of the situation, just as one comes out of a long fainting fit and says, "Where am I?" The mayor is doing everything in his power to straighten matters out.

I ran along in his tracks for three of four blocks this morning and heard him refuse licenses for carts and passes to at least a dozen men within a breath. He threatened a large and healthy-looking colored man with instant death if he didn't stop begging and get to work and help clean up

VIEW OF GALVESTON.

the city. The colored man turned green and gray, but before he could draw his breath to expostulate the chief was gone.

He clutched up three or four men and five or six women and made them race along the street with him to a relief bureau, wrote them orders for food, and would not listen to a word of thanks or explanation. I like the chief of police of Galveston. He knows his business, and he does not care a thing who likes what he does or doesn't like it. He is really the force behind the fine organization which is gradually growing into useful life here now in the reconstruction of the city.

The little parks are full of homeless people; the prairies around Galveston are dotted with little camp-fires, where the homeless and destitute are trying to gather their scattered families together and find out who among them are dead and who are living. There are thousands and thousands of families in Galveston today without food or properties or a place to lay their heads.

It will take thousands and thousands of dollars to put them on their feet again. I believe the people of America will see that money is not lacking. Oh, but, in pity's name—in America's name—do not delay one single instant. Send this help quickly or it will be too late.—Winifred Black in the New York Journal.

CHAPTER VII.

THE PEOPLE IN DESPAIR.

THE ENORMITY OF THEIR LOSSES, WITH SHOCK AND STRAIN, DRIVE MANY TO THE VERGE OF INSANITY.

SEARCHING FOR THE DEAD.

WITH TEARLESS EYES AND FEVERED BRAIN THE PEOPLE GROPE AMONG THE RUINS FOR THOSE THAT CANNOT BE FOUND.

THOUSANDS FLEE FROM THE CITY.

For two days after the great catastrophe the people of the city of Galveston were stunned. They seemed to be dazed. It is a remarkable thing that there were no signs of outward grief in the way of tears and groans to mark the misery that raged in the breasts of the people. Only when some person who was thought to have been dead appeared to a relative living, who had mourned for him or her, were there any tears. There was a callousness about all this that attracted the attention of those who had just come to the unfortunate place. There was a stoicism in it, but it was unexplainable. It indicated no lack of appreciation of what had occurred. It demonstrated no lack of affection for those who had gone.

Nature, generous in this instance, came to their relief in a way and made them dull to the seriousness of what had occurred to an extent which prevented them from becoming maniacs, for if the grief which comes to a mortal when he loses a dead one had come to this whole community the island would have been filled with raving maniacs. In case of individual losses there is always some one near to give consolation. Had the grief come to the whole island there could have been no consolation, for every soul on it had lost in some way that which was dear to it.

"The case is just like the after-thoughts of those who have participated in a great battle," said an old soldier to me. "If a popular man was lost on the picket line there were tears for him, but when the time came for all to be mowed down the horror of it dulled the sensibilities of those who survived."

I was talking to an estimable and bright woman on the subject. She had lost members of her family, though not immediate ones. She said to me: "I study myself and am overcome at myself. I know what has happened. I know the losses. I have lost some of the members of my family, though they are not blood kin. I have lost the dearest friends of my life. And yet I have not shed a tear. My eyes are hot. I would give anything to cry, but it looks as if the fountains were dried. I am ashamed of my seeming indifference to this horrible thing and the loss of those who were so dear to me. But I cannot cry. I know that I suffer, but it looks so cruel to sit here with dry eyes and without any other evidence of the deep sorrow that fills my bosom."

I talked to one man, and asked him how many people he had lost. He had saved his daughter and her child. All the rest, amounting to three souls, were gone. But his eyes were dry. He spoke in a low tone. But it did not tremble. He was agonized—I saw that—but his mind was unable to grasp the true meaning of his loss, and when he had finished he asked if I had a match about me.

Writing on Wednesday night a correspondent of the Dallas News said:

"It has been a day of anguish, like all the days of this week have been. There has been no cessation of tear-stained faces appearing here and there to tell of the lost. And it is a wonder if the end of this sad divulgence will ever come. A motherless boy or a fatherless girl, a now childless mother or father, or whatever it may be, they still come to tell their woe, and the stolid men who glide over the water or who search the shore still bring in the swollen and unrecognizable victims of the storm. It will end some day, and agonizing hearts may rest the painful throbbings of this hour. It matters not how great the numbers of the dead, they are numerous enough to shock the sympathies of the world, and they are gone forever. But we fear to look upon the sea, lest some heartless wave shall bring to view the cold, stark form of somebody whom somebody loved. The victims are still growing into larger thousands, and the bereft are still coming in to tell of losses. It is a continued story of anguish and death such as Texas has never known before and will never know again.

"It is needless to repeat the sad discoveries which every day brings forth. It is said that every wave of the sea has its tragedy, and it seems to be true here. In Galveston it has ceased to be anxiety for the dead, but concern for the living. The supreme disaster, with its overwhelming tale of death and destruction, has now abated to lively anxiety for the salvation of the living.

"I have before me this minute four rings. The man who brings them tells me that they were taken from rigid fingers among the 700 who on last Monday were sunk to rest amid the borderless fathoms of the sea. He says they may be the means of identification of three lost ones. No. There can be no identification, but who can tell the tender secrets which these circlets pledged? Identification is impossible deep down among the mysteries of the sea.

"The tragedy grows greater every moment. The romances dead to the world, the grief lost beneath the wave or carried to the vapors above the earth, the aching hearts soothed by lasting peace, the tired souls in the arms of endless rest, the ambitions stilled by the calm which banishes the anguish of life's dreary struggle—it matters not what these rings may bring to mind—we are yet confronted with the loss of the thousands who shall never again press these wave-kissed shores. The sentiment of this people is, God rest everyone who sleeps beneath the waves, and gather to everlasting peace the ashes of all whose funeral pyres were built of these shattered homes."

The correspondent tried to glean some information from those who had experienced its awful results relative to the storm and its causes. But the stricken people of Galveston could not tell anything about it. They were not agreed even upon the direction of the wind. In their dazed condition memory and reason seemed to have forsaken them. On these points he wrote:

"In years to come men may be able to talk of this greatest of catastrophes in the cool, deliberate way which will admit of reasonable hypotheses as to the causes of the results. But they cannot do it now. The wind blew from the east. The currents were criss-cross. My God! it was awful. And that is as far as you can get with any of those left, for they know no more. They know that the wind blew. They know that the waves rolled. They know, or the most of them do, that they lost dear ones, and that is all. The hydrographer of the future may tell us all. But as far as such people of South Texas as I am, they will leave it to him. He may know the currents and the winds, and tell to the satisfaction of all. But he will never tell of these horrors. I cannot in the present. I may not be able to do it in the future. When the story of the funeral pyres and the burials at sea and the reasons for both are explained—when the pictures are given of the rescued, hunting for the dead—then, indeed, if all are drawn as they are—natural and unstained—another monstrosity in newspaper life will have arisen.

"No man, scientist or mere citizen, is authority upon the wondrous winds and tides that reduced the island of Galveston to an incomprehensible pot pourri of devastation. All is guesswork, behind which there is neither science nor common sense. As a deliberate proposition evolved by a fair measure of judgment, in which there enters as little of egotism as is possible with human beings, I would rather trust the guesser than the scientist."

One of the saddest features of the days that followed the storm was the hopeless search of the living for the dead. There were few among the survivors who had not lost a near relative. In maddening grief, husbands were searching for their wives and wives for husbands. Parents peered amid the ruins for their children, and looked upon the pale faces of the dead in anxious expectancy to find the loved one.

In a few cases the bodies of those sought were found, and in the grief which followed long vigils and the strain of searching nature would collapse, and the living had to be borne away to extemporized hospitals to recover from the shock. In most instances the search proved fruitless, and after being kept up for days was sadly abandoned as hopeless.

The story is told of one man who searched all about the wrecks of houses near where his home had stood for his missing wife and child. Day after day he looked everywhere, constantly widening the radius of the circle over which he wandered. Finally, toward the end of the week he caught sight of a piece of torn cloth which formed part of a woman's dress. He tore away the boards which covered a corpse, and there found the body of his wife clasping the child in her arms.

Gazing upon his lost ones for awhile, the man stood mute and with dry eyes. His grief had passed the point of outward manifestation. He had known for days that they were dead, and he wanted the mournful satisfaction of looking upon their mortal remains. He summoned helpers from those clearing away the debris near at hand, and aided them in carrying the bodies to the fire, into which they were thrown to be cremated. He turned away and joined the workers in another part of the city.

Even little children appeared in that despairing throng who searched for missing friends.

A little fellow straggled in with a crowd of refugees late last night. His name is Frank Caulk. He was all alone. His little face was as white as death but his great eyes blazed courage and a kind of desperate resolution at every glance. He was not hungry, he said; not at all. He was tired, very tired, and he would be very glad to have a place to sleep.

"Anywhere will do for me," he said. "I'll just curl up on the floor." He was thirsty and he was almost naked. I saw him lying in his cot after the nurses had put him to bed. He was looking at some children whose mother was feeding them. His eyes were wolfish.

"See here, my boy," said I, "what is this you are telling me about not being hungry? You are the first boy I ever saw in my life who didn't want something to eat when he could

NEAR 16TH AND K.

get it. Come, now, what is it going to be—some soup, a cup of milk and bread or a nice piece of steak with bread and gravy?"

The boy's great eyes filled with sudden tears. "I—I," he stammered, "I haven't any money; I can't pay for anything."

In about ten minutes he had about hidden his supper, and he ate it like a starving animal. Then I sat down beside him and took his stubby hands in mine and put the coverlet where none could see, and he told me all about it. He had been up in a tree all night the night of the storm, he and his big brother, and in the morning when he looked down his big brother was gone. The water went down inch by inch. He could just see it sinking before his very eyes, but he was afraid to go down again, so he stayed up in the tree all day and all night again. On the second morning he climbed down and found his way somehow to Texas City.

"I was kind of tired," he said, "and something seemed to be the matter with my knees. Every time I'd stumble over anything they would kind o' tremble like. I could not get anything to eat, and my brother was gone. I was afraid to look at the people I found on the sand for fear it would be my brother, and then I met a woman with two children, and she said she was coming up here to you all, and she gave me something to eat, and we slept out on the prairie that night, and this morning we came to Houston.

"My kin are all dead, all but brother. A man told me on the train that brother was alive and was looking for me. I wish he would find me. He was awful good to me that night, brother was. Kept hollering to me not to get scared and to hang on tight. I am so glad he's going to find me, so glad he's going to find me."

He said the last word between a sigh and a yawn, and before he had finished his sentence he was fast asleep. His long lashes made a dark hollow of his great eyes, and his face was like the face of a sleeping baby for peace and innocence.

Having lost their families, some men looked about them for purchasers of their lots where once their homes had stood. They offered them for a song, with the thought that the few dollars they would bring would enable them to turn their backs upon the desolation and seek homes and livelihood elsewhere. In vain their friends counseled patience, saying that the city would rise again. Where they could sell many did so, and hastened away from the scene of their troubles.

As soon as transportation facilities were established thousands fled in despair from the city to friends in Texas or other States. Houston was the mecca toward which they turned. About 1000 were carried up to that city in the two days after the trains began to run. Aboard those trains were men, bareheaded and barefooted, with swollen feet and bruised bodies. Women of wealth and refinement, hatless. shoeless, with gowns in shreds, were among the refugees.

The Houston News of September 18 spoke of one train which came into the Grand Central Depot at 2 o'clock in the morning. It was five hours or more behind time. It had been kept out by delay in boat transportation across to

Texas City from Galveston. There were ten coaches in the train, but they were full. The passengers were chiefly children and women, sick and tired, some injured.

A majority of the passengers who came in on that train wanted to pass through the city, and they were taken over to the Grand Central Depot and transportation arranged for the sufferers to their respective points of destination.

And so for days the trains bore the sorrowing, crushed and despairing away from Galveston. And it is likely that the majority of them will never return. To other towns and cities of the land they will carry the memory of that awful scourge in Galveston, which swept away their possessions and in most instances their loved ones. The majority of the refugees were saddened and broken-hearted men and women, who could no longer endure the sight of the wave-swept island, where the hopes of life lay crushed in the vast heaps of debris or ascended in the smoke of unnumbered funeral pyres. It was the remnant of a stricken people fleeing in despair.

The loss and sorrow that had fallen on so many families was accentuated in a peculiar way by the publication, a week afterwards, of the society column that had been prepared the day before the storm for the Galveston News. To each paragraph in the report as first written was added a few words in parentheses telling what fate had been visited upon the persons mentioned.

In nearly every paragraph there was tragedy. Belles and

beaux whose gay doings were written about a week ago had their obituaries printed with the social news items.

The society reporter was at her house a week before when the storm began, preparing the usual half-page of social news for the Sunday paper. As she wrote the storm grew, and she was driven from the lower floor, where the water was knee-deep, to the upper floor, where she continued to write. Late at night she put aside her copy with the sad consciousness that the city was wrecked and the newspaper office with it. She escaped the flood, but returned home to find the water-stained and torn manuscript written a week before.

One item told in detail of a picnic at the Catholic Orphan Home grounds the night before the disaster, at which all the younger members of society were present. Added in parentheses was this: "The spot now makes one shudder. Ninety people were lost from that one house."

Another item ran that "Mr. and Mrs. W. F. Beers leave this week for New York city, where Mr. Beers will resume his law course at Columbia." In parentheses it is added that the Beers house has been converted into a hospital.

Another item told of a gay tally-ho party to Techman's on the beach the night before the storm. The addition in parentheses said that not a sign of Techman's was left, and that many of the merry guests of the party were swept to death by the storm.

There was the story of a dance on Friday night at the palatial Lobit House, and the addition told of the awful scenes on the night following, when sixty terrified people sought shelter in the same house.

Another item says that "Mr. Ed Ketchum returned from Chicago Tuesday," and adds, "returned to prove himself a faithful soldier of humanity."

The tragedy stunned the living in Galveston, and for days they walked as in a dream, little realizing what was going on about them. There was that terrible sense of loss and disaster ever before their minds, giving sleepless nights and bringing them ever nearer to the verge of insanity. Some recovered sooner than others, but the actual awakening came about Friday morning, when hope seemed to conquer despair in the hearts of that suffering people.

Up to Thursday night there had been no sleep in the city. True, exhausted nature had thrown men and women and children on their beds, and they had closed their eyes, and the physical strain had been to some degree relieved, but the mental strain was still at the breaking point. One man told me that on Thursday morning he was awakened by the convent bell summoning the living to mass. It was the first bell that had rung or tinkled in the town since the day of the storm. He was not a Catholic, but he said it was the sweetest music he ever heard. He bounded from his bed a new man. He was hopeless the day before. He had seriously thought of abandoning his house, which he believed beyond repair, but when he looked at it on Thursday morning it did not look so badly. He resolved to fight it out. He went and found others like himself—resolved to fight it out.

CHAPTER VIII.

THE WORK OF RELIEF.

GENEROUS AMERICA RESPONDS PROMPTLY TO THE CALL OF GALVESTON'S AWFUL NEED.

CHARITY FINDS FIT EXPRESSION.

TRADE ORGANIZATIONS, CIVIC SOCIETIES, NEWSPAPERS, POLICE, CHURCHES, THEATERS AND PRIVATE INDIVIDUALS UNITE IN CONTRIBUTIONS OF MONEY AND SUPPLIES.

The work of relieving the destitute and suffering in Galveston began the day after the storm, and proceeded day by day in all parts of the country. Response was quick and generous to the unusual demand created by the flood. The hungry were fed and the naked clothed with that promptness characteristic of the push of the American people.

It will be recalled that the surviving citizens of Galveston met in the water-soaked parlor of the Tremont Hotel to devise such means of immediate relief as were possible. An organization was effected and everything possible done to meet the emergency that day, even to dispatching a messenger to Houston with an appeal for outside aid. But this was hardly necessary, for the very first news of the disaster was an appeal to generous America. In every city of the land men were on the point of organizing a relief fund even before the messages asking help came from the officials of Galveston and the authorities of the State of Texas.

For two days Galveston had to face the situation singlehanded. Houston was aroused, and special relief trains were sent out on Monday to try and reach Texas City. But the storm had done so much damage to tracks that it was Tuesday before communication was established, and then only imperfectly. But by that time supplies were taken to Galveston and distributed to the hungry people.

On Tuesday, also, the following appeal for help was sent by the Galveston Relief Committee:

"A conservative estimate of the loss of life is that it will reach 3000. At least 5000 families are shelterless and wholly destitute. The entire remainder of the population is suffering in greater or less degree. Not a single church, school or charitable institution, of which Galveston had so many, is left intact. Not a building escaped damage, and half the whole number were entirely obliterated. There is immediate need for food, clothing and household goods of all kinds. If nearby cities will open asylums for women and children the situation will be greatly relieved. Coast cities should send us water as well as provisions, including kerosene oil, gasoline and candles. W. C. Jones, mayor; M. Lasker, president Island City Savings Bank; J. D. Skinner, president Cotton Exchange; C. H. McMaster, for Chamber of Commerce;

R. G. Lowe, manager Galveston News; Clarence Owsley, manager Galveston Tribune."

On that day, also, General Spaulding, acting Secretary of the United States Treasury, received the following official statement of the situation from Postmaster Griffin and Special Deputy Collector Rosenthal at Galveston:
"Secretary of the Treasury, Washington:

"The city and island of Galveston swept by terrific cyclone and tidal wave of unprecedented fury. The entire city is inundated, and the Gulf encroaches for several blocks. The residence part is in ruins and many people homeless. The dead, it is feared, will reach about 1500, and perhaps twice as many. Streets are obstructed by debris and dead animals, and wires are in every part of the city. There is more than eight feet of water in stores and warehouses, damaging stocks of goods and provisions. Thousands of people are homeless and wounded, and some 500 are sheltered in the custom-house, which is practically roofless.

"All railway communication is shut off, and wagons and railway bridges leading to the mainland are gone. Ocean steamers to the number of seven or eight swept ashore and small craft demolished. Life-saving station swept away and no trace of crew. Lightship up in West bay; occupants supposed to be safe. Old custom-house roofless and windows blown out. All stored merchandise, principally sugar, badly damaged. Boarding boats swept away, and barge office badly wrecked. Need tents and 30,000 rations.

"Citizens' Relief Committee doing all in their power, but

VOLUNTEERS REMOVING DEBRIS TO OPEN STREET, UNDER GUARD, TWENTY-FIRST LOOKING NORTH.

CORNER STRAND AND TWENTY-SECOND STREET, LOOKING WEST—ONE STORY BLOWN OFF MOODY BANK BUILDING.

stock of undamaged provisions is exhausted. With all the people housed in building, will need extra force of six men to keep building in sanitary condition. Have hired boat to take dispatch to mainland for transmission. Relief urgently requested."

These appeals, with one sent out by the governor of the State, gave official notification to the people of the United States that large funds were needed in the ruined storm district. It was enough. In every city of the land the municipal officials, trade and civic organizations, newspapers, the police departments, the churches and private charity did something for the sufferers from the Galveston flood.

There were two batteries of United States artillery in Galveston, one on either side of the channel, commanded by Capt. William C. Rafferty. The barracks were destroyed, the books of record and most of the belongings of the companies, but nearly all the men were saved. As soon as communication was established news of the disaster was sent to the War Department at Washington. As soon as the Western Union Telegraph could forward the orders the machinery of the army was set in motion to relieve the storm sufferers.

General Weston, chief of the commissary department of the War Department, on September 12 received word from the officer in charge of Fort Sam Houston, at San Antonio, 240 miles from Galveston, that 20,000 rations had been shipped.

An additional 30,000 rations were shipped from St. Louis

the same day by order of the War Department. An army ration consists of four pounds of water-free food, sufficient for three meals.

About 1500 tents were shipped to Galveston from Fort Myer, the Arsenal in Washington and other nearby posts.

Acting Secretary Meiklejohn wired to St. Louis directing that a special train be despatched to Houston at once with supplies.

By direction of President McKinley, General Spaulding, acting Secretary of the Treasury, directed the revenue cutter Winona, then at Mobile, and the lighthouse tender Arbutus, at New Orleans, to proceed to Galveston at once. They were to report to the collector of customs there for such duty as Governor Sayers of Texas may direct. Governor Sayers had pointed out the need of a vessel to ply between Galveston and the mainland, practically all sailing craft having been put out of service by the storm. Both of the vessels were to do duty of that kind, if needed.

The Navy Department sent the gunboat Bancroft from New London, Conn., and the tug Samoset from Pensacola, Fla. Both were ordered to co-operate with the municipal and State authorities in whatever service should be required.

One of the first dispatches to reach Washington after communication with Galveston was opened through Houston was the following to the quartermaster-general:

"Referring to my telegrams of 9th and 10th, I have, subject to approval, suspended Fort Crockett construction contracts, and again urgently recommend that contractors be paid for labor and material in places and on ground all swept away and lost beyond recovery. Fortifications at Crockett, Jacinto and Travis all destroyed and can't be rebuilt on present sites. Recommend continuance of my office here only long enough to recover Crockett office safes and morning gun, when located; also to close accounts and ship my office and recovered property where directed.

"I fear Galveston is destroyed beyond its ability to recover. Loss of life and property appalling.

"BAXTER,
"Quartermaster."

The three forts mentioned in the above dispatch were situated so as to command the harbor of Galveston. Fort San Jacinto was on the northwest end of Galveston Island. Across the channel leading into Galveston bay, between the island and the mainland and north of Jacinto, is Bolivar Point, upon which was situated Fort Travis. At the southeastern end of the island, upon which Galveston is situated, was Fort Crockett.

Adjutant-General Corbin also telegraphed instructions to General McKibbin, commanding the Department of Texas, at San Antonio, to proceed to Galveston at once and investigate the character and extent of the damage caused by the recent hurricane and to report to the Secretary of War what steps are necessary to alleviate the sufferings of the people and improve the situation.

Upon his arrival on September 12 General McKibbin informed the War Department as follows:

"Arrived at Galveston at 6 P. M., having been ferried across bay in yawl-boat. It is impossible to adequately describe the condition existing. The storm began about 9 A. M. on Saturday, and continued, with constantly-increasing violence, until after midnight. The island was inundated; the height of the tide was from eleven to thirteen feet. The wind was a cyclone. With few exceptions, every building in the city is injured. Hundreds are entirely destroyed.

"All the fortifications except the rapid-fire battery at San Jacinto are practically destroyed. At San Jacinto every building except the quarantine station has been swept away. Battery O, First Artillery, lost twenty-eight men. The officers and their families were all saved. Three members of the hospital corps lost. Names will be sent as soon as possible. Loss of life on the island is possibly more than 1000. All bridges are gone, water-works destroyed, and all telegraph lines are down. Colonel Roberts was in the city, and made every effort to get telegrams through. The city is under control of Committee of Safety, and is perfectly quiet. Every article of equipment or property pertaining to Battery O was lost.

"Not a record of any kind is left. The men saved have nothing but the clothing on their persons. Nearly all are without shoes or clothing other than their shirts and trousers. Clothing necessary has been purchased, and temporary arrangements made for food and shelter.

"There are probably 5000 citizens homeless and absolutely destitute, who must be clothed, sheltered and fed. Have ordered 20,000 rations and tents for 1000 from Sam Houston; have wired commissary-general to ship 30,000 rations by express. Lieutenant Perry will make his way back to Houston and send this telegram.

"McKibbin."

On the same date Capt. Charles Rich, the engineer officer in charge of river and harbor improvement at Galveston, sent this dispatch to General Wilson, chief of engineers:

"Hurricane caused tide twelve to fourteen feet above mean low. Jetties seem to have settled; cannot yet be seen with tide three feet above; probably seriously damaged. Batteries practically ruined; nothing but concrete portions left. Casemate torpedo buildings, warehouses, coal wharf wiped out. Dredge Comstock beached on Pelican flats; will have to be dug out; crew all saved. Tug Anna at Velasco, in Brazos river, has not yet been heard from. Assistant Engineer Tallfor was on the Anna; Superintendent Campbell probably also. Superintendent Hinkle at Aransas Pass; not yet heard from. Other assistants and clerks safe. Captain Judson's wife reported here drowned; cannot verify this. Self and family safe. Rafferty, Baxter, Nichols and Longive safe; they have probably wired department already.

"Tug Anna ashore in Brazos river, just below lighthouse; all hands safe. Depth of water in usual entrance channel here reported shoaled to twenty feet; probably is deeper elsewhere. Have arranged with Contractor Clarke to float Anna; also to bring his dredge from Plaquemine bayou here to dredge out government dredge Comstock, so she may be available for urgent work."

While the War Department was thus making investigations regarding its property and men and doing what it could to relieve the sufferers, Gov. Joseph D. Sayers, governor of Texas, and the mayor of Galveston and the Citizens' Relief Committee were giving their official attention to the work of getting supplies to the stricken city as soon as possible. One of the first acts of Governor Sayers was to issue an order placing three regiments of the militia at the disposal of Gen. Thomas Scurry and to send him immediately to the city to aid the citizens in maintaining order.

On Monday after the storm a dispatch was sent to President McKinley by the Citizens' Relief Committee:

"Houston, Texas, September 10, 1900.
"William McKinley,
    "President of the United States,
        "Washington, D. C.:

"I have been deputized by the mayor and Citizens' Committee of Galveston to inform you that the city of Galveston is in ruins, and certainly many hundreds, if not thousands, are dead. The tragedy is one of the most frightful in recent times. Help must be given by the State and nation, or the suffering will be appalling. Food, clothing and money will be needed at once.

"The whole south side of the city for three blocks in from the Gulf is swept clear of every building. The whole wharf front is a wreck, and but few houses in the city are really habitable. The water supply is cut off, and the food stock damaged by salt water. All bridges are washed away, and stranded steamers litter the bay. When I left this morning the search for bodies had begun; corpses were everywhere. Tempest blew eighty-four miles an hour, and then carried government instrument away; at same time waters of Gulf were over whole city, having risen twelve feet. Water has now subsided, and the survivors are left helpless amid the wreckage, cut off from the world except by boat.

"(Signed) RICHARD SPILLANE."

In reply to this appeal the President sent the following:

"Washington, D. C., September 10, 1900.

"Hon. J. D. Sayers,
"Governor of Texas,
"Austin, Texas:

"The reports of the great calamity which has befallen Galveston and other points on the coast of Texas excite my profound sympathy for the sufferers, as they will stir the hearts of the whole country. Whatever help it is possible to give shall be gladly extended. Have directed the Secretary of War to supply rations and tents upon your request.

"(Signed) WILLIAM MCKINLEY."

The publicity given to this official news of the needs of Galveston aroused the charity of the whole country. There were many large individual contributions.

George McFadden, the noted cotton operator, cabled £1000 for the relief of the distressed on the day after the storm.

The mayor of Houston sent out telegrams to the mayors of the large cities soliciting their co-operation in securing funds. The message to Mayor Van Wyck of New York brought the following reply, and he headed the New York subscription with his pledge of $500:
"Hon. S. E. Brashear,
 "Mayor Houston, Texas:
"In response to your telegram I have issued a call to the people of the city of New York to contribute to the relief of those afflicted by the disaster at Galveston. Please express to the mayor of Galveston the profound sympathy of the people of New York for the people of Galveston in this hour of their distress.
"ROBERT A. VAN WYCK."

In response to a telegram from Miss Barton, Mr. Flather recently sent $1000 from the fund in his charge. Acknowledgment has been received.

Joseph Seligman of J. W. Seligman & Co. of New York contributed $1000.

Woodward & Lothrop of Washington gave $200.

Robert B. Easley, local manager of R. G. Dun & Co., received a telegram from the main office in New York instructing him to authorize the local manager in Galveston to draw for $1000 to make himself and office force comfortable.

The Paris Medicine Co. of St. Louis, Mo., sent to Galveston $200 worth of medicines and $100 cash.

The Christian Herald of New York gave $1000, and was among the first contributors.

Copyrighted by Leslie's Weekly, 1900. *Drawn by E. Johnson.*

NO MERCY FOR GHOULISH LOOTERS AT GALVESTON.

Twelve firms in Washington gave each $100 to the subscription.

Robert Garrett & Sons, bankers, of Baltimore, contributed $1000.

One of the subscribers to the Galveston relief fund in Boston was Joseph Jefferson, the veteran actor, who gave his personal check for $1000. Mr. Jefferson's gift was inspired by reasons of affection as well as by sympathy. He attended school in Galveston, and his earlier successes in his dramatic career were gained in that city.

Three little girls of Baltimore sent $1.50, the proceeds of a lemonade stand, kept open one day and evening.

One old lady in Philadelphia brought $1 to the police captain in whose precinct she lives, saying that it was all she had.

In one week the various agencies, as reported by the New York Times of that city, raised $208,081.40.

The Citizens' Relief Committee of the Merchants' Association of New York raised $70,000, and loaded the U. S. transport McPherson, loaned by the government for the purpose, with supplies direct to Galveston. Part of the cargo was thus enumerated:

Two thousand barrels flour, 10,000 bags charcoal, 3000 gasoline stoves, 1000 barrels copperas, 500 barrels chloride of lime, 200 barrels solution of carbolic acid, 5000 half-barrels cornmeal, 5000 bags rice, 5000 bags white beans, 1000 barrels split peas, 1000 drums codfish, 750 single sacks roasted Mocha and Java coffee beans, 25 chests of tea, 100 barrels granulated sugar, 1000 tins baking powder and 1000 pails lard.

Chicago, not to be outdone, chartered a train and sent it loaded to Texas City. The train was made up of five cars.

No less than four newspaper relief trains were sent, carrying large supplies.

The Denver Republican started a subscription with $100, and a large fund was raised in that city.

The Cleveland Chamber of Commerce raised a fund which amounted to $3000 in two days.

Richmond, Va., subscribed a fund of $6000.

Norfolk was only a little behind, with $4890.

In Washington a game of baseball was played at the National Park, and the gate receipts were given to the fund. In the game one of the players received injuries from which he died four days afterward.

In many of the large cities the theatrical troups and local managers gave the proceeds of one day and night to the relief fund of the city in which they were playing. In Boston, New York, Philadelphia, Washington, New Orleans, St. Louis, Cincinnati, Chicago and Denver thousands of dollars were raised in this way. In Baltimore the police sold $5000 worth of tickets to the several performances of the day and night in that city.

A rough estimate of contributions from various States and cities made up of newspaper reports is as follow

STATES.

	Am't raised.	Total expected.
Texas	$350,000	$500,000
Alabama	13,000	18,000
Pennsylvania	150,000	300,000
District of Columbia	7,000	15,000
Wisconsin	15,000	100,000
Minnesota	21,000	50,000
North Carolina	4,500	13,500
Kentucky	12,735	15,000
Virginia	25,000	35,000
Indiana	7,000	14,000
Kansas	30,000	50,000
Florida	5,000	10,000
Louisiana	40,000	50,000
Nebraska	6,242	10,000

CITIES.

	Am't raised.	Total expected.
New York	$250,000	$400,000
Atlanta	7,117	12,000
Buffalo	6,000	9,000
Milwaukee	12,000	20,000
Baltimore	13,500	28,500
Philadelphia	65,000	200,000
Montgomery	4,000	10,000
Charleston	6,400	7,400
Syracuse	894	1,700
Boston	33,620	75,000
St. Louis	69,453	150,000
Detroit	4,298	7,000
Chicago	79,600	160,000
Little Rock	3,000	5,000
Louisville	11,110	20,000

Lexington.......................	1,200	2,000
Providence.....................	5,592	10,000
Rochester......................	2,720	3,500
Utica..........................	950	10,000
Indianapolis...................	3,075	5,000
Albany.........................	618	2,000
Johnstown......................	4,000	5,000
Jacksonville, Fla..............	1,500	3,000
Tampa..........................	1,700	4,000
New Orleans....................	35,000	50,000
San Francisco..................	15,000	50,000
Denver.........................	12,604	20,000
Omaha..........................	3,532	5,000
Washington.....................	8,464	12,300

The total contributions are estimated at more than $1,500,000.

Churches and religious organizations raised large contributions, and the organized societies also were active all along the line. Local Masons, Odd Fellows, Knights of Pythias and other societies became so many centers of aid through their affiliations throughout the Union.

Even foreign countries responded. The following message of condolence was sent by Emperor William III of Germany to the President and by him to Governor Sayers:

"I wish to convey to your Excellency the expression of my deep-felt sympathy with the misfortune that has befallen the town and harbor of Galveston and many other parts of the coast, and I mourn with you and the people of the United States over the terrible loss of life and property caused by the hurricane; but the magnitude of the disaster is equaled by the undauntable spirit of the citizens of the world, who, in their long and continued struggle with adverse forces of nature, have proved themselves to be victorious. I sincerely hope that Galveston will rise again to new prosperity.

"WILLIAM, I. R."

The President of France also sent a message of condolence:

"The news of the disaster which has just devastated the State of Texas deeply moves me. The sentiments of traditional friendship which unite the two republics can leave no doubt in your mind concerning the very sincere share that the President, the government of the republic and the whole nation take in the calamity that has proved such a cruel ordeal for so many families in the United States. It is natural that France should participate in the sadness as well as in the joy of the American people. I take it to heart to tender to your Excellency our most heartfelt condolences, and to send to the families of the victims the expression of our afflicted sympathy.

"EMILE LOUBET."

Similar expressions of good will were received from Great Britain, Italy, Russia and Belgium. Americans in Paris raised and sent to Governor Sayers $10,000 for the sufferers.

A meeting of Americans was held at the United States embassy in Berlin in aid of the Texas sufferers. Resolutions of sympathy were passed, and 2000 marks were subscribed. A committee, consisting of Secretary of Embassy Jackson, the Rev. Dr. Dickie and Deputy Consul-General Frederick von Versen, was appointed to solicit further subscriptions.

Miss Clara Barton, president of the Red Cross Society, was in Washington at the time of the storm. She immediately set about going at once to Galveston to aid in taking care of the sufferers. She issued the following appeal to the people of the United States, and, with Ellen S. Mussey, started for the stricken city:

"The American National Red Cross at Washington, D. C., is appealed to on all sides for help and for the privilege to help in the terrible disaster which has befallen Southern and Central Texas. It remembers the floods of the Ohio and Mississippi, of Johnstown and Port Royal, with their thousands of dead, and months of suffering and needed relief, and turns confidently to the people of the United States, whose sympathy has never failed to help provide the relief that is asked of it now. Nineteen years of experience on nearly as many fields render the obligations of the Red Cross all the greater. The people have long learned its work, and it must again open its accustomed revenues for their charities. It does not beseech them to give, for their sympathies are as deep and their humanity as great as its own, but it pledges to them faithful old-time Red Cross relief work among the stricken victims of these terrible fields of suffering and death.

" 'He gives twice who gives quickly.'

"Contributions may be wired or sent by mail to our treasurer, William J. Flather, assistant cashier Riggs National Bank, Washington, D. C.; also to the local Red Cross committee of the Red Cross India Famine Fund at 150 Fifth avenue, New York city, and the Louisiana Red Cross of New Orleans, both of whom will report all donations for immediate acknowledgment by us.

"CLARA BARTON,
"President American National Red Cross."

The Red Cross was welcomed in Galveston, and soon found enough to do. On September 18, two days after her arrival, Miss Barton sent out a second appeal:

"Find greatest immediate needs here are surgical dressings, usual medicines and delicacies for the sick. No epidemic, but many people are worn out with suffering and exertion, who need careful care and proper food.

"CLARA BARTON."

The Red Cross fund placed at the disposal of Miss Barton was about $3000 up to September 20.

On September 11 the coming of supplies was reported and the situation described in the press dispatches:

"Today supplies began to arrive, but so meagre are the facilities that the amount was really pitiable. The food that got in came from Houston in steamers, but there are so many to feed that it did not go far. The provisions that came were heartily welcome. A crowd quickly gathered, and they were distributed in a short time. Some cooked and ate the food where they stood.

"There is much necessary work to be done, the first thing being the establishment of a rigid system of issuing supplies. The nucleus has already been formed, and the regular soldiers who are still alive and a number of citizens have been sworn as special policemen. These are supervising the issuing of rations, and are also directing the efforts of the searchers of the dead and injured. The provisions that are still in the wholesale and retail groceries have been searched out, and there is quite a quantity of good stuff.

"As for clothing, many have the money to buy, and those who have not are being given their necessaries by the relief committee. But more clothing is necessary, especially women's and children's. Women and children are the principal sufferers now."

On September 12, from the same source, it was learned that the relief work now under full sway at Houston is along two lines—to succor those who cannot leave Galveston, and to bring out of the city all of those who can and are willing to leave.

Mayor Jones and the Citizens' Committee of the Island City are urging that only those shall be permitted to enter Galveston whose presence is imperative, and transportation lines are straining every nerve in order that they may accord the privilege to those who are pleading to get away from the scenes of horror and desolation around them.

On September 13 the correspondent of the Dallas News wrote to his paper:

"The relief system is fairly in operation, and it is now apparent that no one need go hungry, except able-bodied men who refuse to labor. But it should be understood that those desiring relief should go to the different ward headquarters

SACRED HEART CHURCH (JESUITS). THIRTEEN BLOCKS FROM EAST BEACH. ONLY ONE HOUSE LEFT.

CITY RAILROAD COMPANY POWER HOUSE, LOOKING WEST.

TWENTY-NINTH AND MARKET—TAYLOR COMPRESS.

MALLORY WHARF, S. S. ALAMO.

or send some one. The committees and heads of departments have no facilities for forwarding goods to the destitute in the various portions of the city. Their time is taken with procuring and distributing supplies from the various headquarters."

Near the end of the week he described matters as follows:

"The stores and groceries are again getting down to business, but they are badly handicapped by damaged stock, more especially the dry goods and clothing stores. A complete overhauling of these establishments has been necessary, and the separation and sorting out and drying out of damaged goods is not yet completed. Those which have fully opened for business are crowded with customers, and in some instances it is still necessary to keep the crowds out, letting in only a few customers at a time. The clerks are a hard-worked set of people just at the present time. With the changes in overhauling the stock, they have not yet become acquainted with the exact location of articles called for, and it requires a short search to find them. This naturally retards the quick execution of business and throws additional labor on clerks waiting upon customers. But order is rapidly being made out of the chaos existing after the storm, and in course of time things will be moving along with their old-time uniformity.

"The street forces have got fairly to work on the business streets, and they are rapidly assuming a more passable condition. Drays are hauling away the trash, and in the course of a week or so the evidences of the storm will be removed.

The damaged buildings will take longer to repair, but the streets will present more of the old-time aspect than for the past week."

After Thursday matters mended. Communication with the outside world visibly improved, supplies came more regularly and in larger volume, and the worst was over.

On September 16 Mayor Jones dictated this telegram to the press:

"The situation in Galveston is growing better each day. The weather since the storm has been all that could be wished for.

"Systematic distribution of supplies is being made by duly-appointed committees. The streets and alleys are being cleaned of debris by local laborers, who are well organized and whose only compensation is their daily food. With the great generosity and sympathy of the world destitution here will be relieved, confidence restored, and Galveston will again take her proud commercial position among the ports of the world.

"Money, clothing and disinfectants are much needed. All remittances should be made to John Sealy, treasurer of the Relief Committee, and supplies to William A. McVitie, chairman of Relief Committee."

W. C. Jones, mayor of Galveston, sent the following dispatch to the Secretary of War, expressing gratitude for the assistance rendered to the city through that arm of the government service:

"Secretary of War,

"Washington, D. C.:

"The people of the city of Galveston desire me to return to you their heartfelt thanks for your assistance in their hour of trouble and affliction.

"(Signed)   W. C. JONES."

R. B. Hawley, representative in Congress from the district in which Galveston is, sent a dispatch to Acting Secretary Meiklejohn of the War Department as follows:

"Your telegram of 12th arrived tonight. The assurance of your complete sympathy and prompt and substantial aid is received by the citizens of Galveston with profound gratitude. The number of dead can only be estimated. Not less than 5000 lives have been sacrificed. It is the event of the century. Two storms of tremendous velocity met directly here. The frightful results have been largely described by the press. General McKibbin is here, and a great aid and comfort in the work of relief. The tents and rations will be of great service. The destruction is so great we have been compelled to make our conditions known and ask for the sympathy and aid of mankind. If you deem expedient, no greater service could be done than to use such agencies as you may for universal information concerning our unhappy fate.

"(Signed)   R. B. HAWLEY."

CHAPTER IX.

GALVESTON'S EARNEST APPEAL.

ASKS FUNDS FOR HOUSES FOR TEN THOUSAND HOMELESS.

COMMITTEE SAYS POVERTY-STRICKEN SHOULD BE RESTORED TO PLANE OF SELF-RESPECT AND SELF-SUPPORT.

An appeal was issued by the relief committee of Galveston September 26. It is signed by Mayor Jones and the other members of the committee and endorsed by Governor Sayers. It is as follows:

"To the American people:

"Seventeen days after the storm at Galveston it is still impossible to accurately estimate the loss of life and property. It is known that the dead in the city will number at least 6000, or approximately one-sixth of the census population. The island and adjacent mainland will add perhaps 2000 to this number. Actual property damage is incalculable in precise terms, but we have the individual losses, and losses in public property, such as paving, water-works, schools, hospitals, churches, etc., will easily amount to $30,000,000. This estimate takes no account of the direct and indirect injury to business. Along the beach front upward of 2600 houses by actual map count were totally destroyed. Of these, not a timber remains upon the original site, and the wreckage constitutes the embankment of debris extending along the entire beach from three or four blocks inward for about three miles. The removal will cost about $750,000 to $1,000,000. From this debris there are still daily uncovered by the workmen now systematically employed from thirty to fifty bodies, which are burned or buried on the spot. Moreover, we estimate that 97½ per cent. of the remaining houses throughout the city were damaged in greater or less degree. On the removal of this debris in the clearing of the streets, to make temporary repairs to houses temporarily destroyed, in distributing supplies and in the general work of restoration, our entire citizenship are engaged.

"Men whose services could not be secured at this season ordinarily are giving their time without compensation. Firms whose affairs ordinarily require the attention of three partners retain one for the transaction of their business and lend the other two to the public service. The stevedores, cotton jammers and other bodies of organized skilled workmen who command handsome wages at this time of the year have been giving their time free of cost, and one association has besides contributed from the charity funds $1000 in cash to the general relief, while all other organizations are caring for their own to the utmost of their resources. This devotion to the general welfare at the expense of private interests is shown by all classes. Visiting newspaper men now here will bear witness to this universal condition of self-sacrifice.

"The burdens of our committees have been lightened in a

great degree by Governor Jos. D. Sayers, who has personally undertaken the supervision of relief to the mainland sections, and by the splendid success of Adjutant-General Scurry and staff in maintaining discipline and directing labor until the beginning of this week, when his department was merged into the regular relief work and city government. We cannot command the language to express our gratitude for the generous supplies of food, clothing, disinfectants, etc., from all quarters and all agencies. Nor can we overrate the service performed by the American press. And we desire to make special recognition of the metropolitan newspapers and the Red Cross Society that have raised funds and sent relief trains and cargoes, and whose representatives are here not only in the distributing of their supplies, but joining hands with us in the sorrowful and strenuous work of the occasion. By the world's generosity there have been no hunger and now no nakedness in Galveston. And especially do we return grateful thanks for the surpassing and unlimited generosity of the railroads, express, telegraph and telephone companies, without whose prompt assistance we would longer have remained in isolation and chaos, and who are not only rendering services for relief work free of charge, but are giving it precedence over any other business.

"The munificent contributions sent to the governor and directly to the relief committee are perhaps sufficient to defray the expenses of removing the wreckage and disposing of the dead bodies and meeting the most urgent sanitary requirements. The homeless will still not be without shelter when this is done.

"Our people are meeting this disaster with characteristic American pluck. While not forgetting their dead, they nevertheless hide their sorrows and turn their faces cheerfully toward the future.

"Were our task but to afford temporary relief and to care for the wounded and the orphans, an appeal to Texas alone would be sufficient. The wounded and orphans are comparatively few, since only the sturdiest were able to combat the maddened elements. But a greater and a graver work confronts us. Some kinds of homes, be they ever so humble, must be provided for the 10,000 people now huddled in ruined houses, public places and improvised camps, to the end that they may not become paupers, but may speedily set up their households wherein repose all that is best and noblest in American life. We believe that the well-to-do and the charitable people of this nation will not be contented to merely appease hunger and bind up bruises, but will in very large measure and with more far-reaching effect contribute to the restoration of this people to a plane of self-support and self-respect. It is for this purpose that we make this further appeal.

"For such temporary measures as are explained in the foregoing we have, at present, sufficient supplies; but they are only a tithe of the larger needs herein set forth, and generous people of the nation will best serve the situation and their own aims by making their contributions in money.

"We refer to Governor Sayers, Miss Clara Barton of the Red Cross or to any prominent firm or bank in Texas in regard to business ability and discretion of our central relief committee.

"In the midst of sorrow such as no other American community ever suffered we are consoled by the gracious sympathy and prompt relief tendered by our countrymen and other nations in the demonstration of kinship of the world. With our last breath we will bless the donors of these funds, and our every effort will be devoted to proving our people and our city worthy of their assistance."

Miss Clara Barton of the Red Cross Society adds the following to the appeal:

"The manly, straightforward statement of Mayor Jones and his assistants, portraying the condition of this stricken city and the needs of its surviving inhabitants, has been courteously passed to me for endorsement of approval. Most heartily do I endorse it, and I vainly search for words with which to fittingly commend the bravery and efficiency with which the business men of Galveston and the stricken vicinity have met the terrible conditions of their misfortune and their capability of wisely, equitably and honorably continuing the work of relief and rehabitation. Could the people of our generous country see as I have seen in its dreadful reality the desolation and the destruction of homes by thousands, the overwhelming bereavement in the loss of near and dear ones, and the utter helplessness that confronts those remaining, the appeal of Mayor Jones for continued help would

SCENE NEAR THE OCEAN FRONT.

meet with such a response as no other calamity has ever known."

From a penitentiary in a Northern State came a contribution for the relief of Galveston sufferers. In the letter which accompanied the $5 a convict wrote to Mayor Jones:

"I hope it may be the means of relieving some unfortunate Texan. I am a native of Texas, and at present confined in this penitentiary as the result of allowing myself to be made the dupe of a scoundrel whom I befriended. But though a convict, I have not lost interest in Texas, and I only wish that I was able to send you a much larger amount. You will oblige me by not printing my name in the list of contributors."

The Mayor's voluminous mail is telling how the calamity reached the hearts of the nation. In one delivery there came letters from Bowling Green, Ky.; Freeport, Kan., and Irvona, Pa., offering homes to Galveston orphans.

"My wife and I would be glad to befriend some homeless, helpless girl who has been bereft by the storm, one without parents or anyone to care for her," wrote the Bowling Green man. "We will give her a home and do all we can to comfort her and bring back the sunshine into her life."

"If you have in your care a little girl whose parents and friends have been killed in the storm I will take her and give her a good home," wrote the Kansas woman, giving as references the leading merchants of Freeport.

"If you could send me two little girls that have lost parents and home I would be pleased to raise them as if they were my own children," the Pennsylvania man said.

These are samples of scores of similar offers from all parts of the country. Galveston orphans will be cared for. One of the earliest arrivals after the storm was Mr. R. C. Buckner, representing the Buckner Orphanage at Dallas. As soon as the children could be gathered a party of twenty-six was taken to the orphanage. Room for seventy more was offered if it should be needed. But perhaps most notable was this proposition from the Buckner institution:

"We can admit twenty crippled, destitute children into the children's hospital. Everything will be free to them. Yours for God and humanity."

A common sight on the outgoing boat was a sister or a priest having in charge a group of little ones en route to some Catholic institution. The Odd Fellows came here to help their brothers in distress, and found that thirty-five members of the order had perished out of a lodge membership of 215. They made prompt provision for the transfer of nine orphans to the home of the order at Corsicana. One of the little folks told the Odd Fellows' relief committee how his baby brother was lost in the flood.

"Mamma," he said, "had baby, holding him with her teeth. She had to let go to get her breath. He dropped in the water and she never could get him again."

The poor mother was actually holding the child with the clothes gripped in her mouth, while she clung to a raft, when a wave dashed over her head. She gasped, and the baby was swept away.

One of these Odd Fellows saved two children out of his large family. When the house went to pieces he swam with the two little girls clinging to him. As he was becoming exhausted he caught a sugar barrel floating by. The children were lifted inside, and the father held it, warding off drift and keeping the barrel from capsizing until the water went down.

There will be no want of provisions for the orphans of Galveston. Dr. Morris, the superintendent of the Texas Children's Home Society, said:

"You will be surprised when I tell you that I have applications for thirty or forty children more than I have, and some of the applicants who wish to adopt are very wealthy."

The relief problem was very much lightened by the action of the insurance companies.

The State agents of most of the life insurance companies prepared a complete list of their respective policy-holders residing in Galveston at the time of the storm, and they announced that every facility would be offered for the making of proofs of death. All claims, it was announced, would be paid promptly, regardless of lost policies.

It was conservatively estimated by these agents that the life insurance in force at Galveston would aggregate $6,000,000. One company had over $1,500,000 in force there. Few claims have yet been presented, and many will probably never be heard of, as whole families and all papers were destroyed. When possible, the companies looked up the nearest relatives and paid the claim to them.

Unusual problems in law and business followed hard upon the Galveston hurricane. On the second day after the banks opened a customer presented himself at the counter of one of them with a difficult question. He had deposits there for which two certificates had been given him. The certificates had gone with the wind and wave. If they should turn up in Calcutta and come through banking channels at some time in the more or less distant future they would have to be paid. The probability was that the negotiable paper had been destroyed. The possibility was that it might be presented at any time. How should the Galveston banks do justice to the holders of certificates and at the same time protect themselves to a reasonable degree against double payment of the negotiable evidences of indebtedness scattered to the four quarters of the earth?

"We shall recognize the great emergency," Vice-President Louis R. Bergeron of the Galveston National Bank said. "It will not do to stand upon the technicalities which apply in ordinary transactions. I suppose some kind of an obligation to make good to the bank the money paid in case the certificate shall be found and presented will be the course. Most depositors had passbooks. If these are lost it will not interfere with business. New books can be issued."

The distribution of life and beneficial insurance will not be attended by the difficulties which were expected. Very promptly the companies having the largest amounts at stake adopted a policy of extreme liberality in the construction of policy provisions. Other life and beneficial organizations followed, until there was none standing out for technical "proofs of death" required in ordinary circumstances. One of the provisions of most policies is that the statement of at least one person who was present at the death must be furnished. This was waived at Galveston, and "reasonable" evidence that death had taken place was sufficient. The instructions given to the life insurance agents were such that the word "reasonable" was interpreted with considerable latitude by them. One company had 1000 policies in Galveston. Of the holders only six are actually known to have died in the storm. Another company which dealt in small policies of $150 or thereabouts, and which collected premiums weekly or monthly, had about 7000 people of Galveston on its list and employed twenty-five agents. This company loses more proportionately than those which issued larger policies. The agent upon his arrival realized it would be utterly impossible to enforce the usual rules of production of policies and proofs of death. These policies provided for payment within twenty-four hours of death. They were designed to provide for funeral expenses, and appealed to people who had little. This company has adopted the plan of accepting testimony which gives reasonable presumption that the policy-holder was killed in the hurricane.

Before the first week had passed the public-spirited citizens who had undertaken the work of relief and restoration realized the problems likely to arise. They immediately established what might be called a legal evidence bureau. The scope of this branch of effort to ameliorate the sad conditions was set forth in the following notice in the newspapers:

"Any and all persons possessing legal evidence of the death of anyone in connection with the recent disaster at Galveston are requested to appear at central relief station in Goggon Building, Market and Twenty-second streets, to make affidavit to same. It is desired to at once collect as much authoritative evidence as possible for future legal reference in matters of inheritance and insurance. Witnesses not now resident in Galveston are requested to send sworn statements of their knowledge as early as possible to Board of Health, mortuary committee, Galveston, including in evidence the name, age, sex, color, occupation and residence of deceased, and at the same time detailing fully upon what basis such statements are founded. All persons subscribing should attach their places of residence to the subscription and indicate other indiviudals able to substantiate their evidence if possible."

The importance and value of the material thus collected was increased as questions of inheritance arose for the courts to pass upon. For the time being all such issues were in abeyance. But there will be not a few cases where heirs in other parts of the country will enter conflicting claims to real estate, which, with the return of confidence, will be held in much greater estimation than it is now. Whole families with considerable estates were destroyed. Such property as remains will go to non-residents.

Galveston's rating as the third richest city, proportionate to population, in the country has been mentioned. Assessment was on a scale of about one-third of the actual value. With the growth of the city the tax figures had not been increased materially. This would give a real valuation before the storm of $75,000,000. Money is already arriving here for investment in realty. Of course, the expectation is to buy at low figures from people who are discouraged and want to get away.

In front of a real estate office at the beginning of the second week was chalked up: "Fifty thousand dollars to invest in Galveston realty at storm prices." Little property has changed hands. One house and lot, the building in fairly good condition and representing a cost of $3500, sold for $1000 in cash, the owner desiring to leave the city. Inside property, that is, the business district, is not on the market at any material reduction from the prices which prevailed before the storm.

Tornado insurance is popular in some parts of Texas where cyclones have furnished object-lessons. Galveston people felt so secure that they carried very few policies of this kind, perhaps a less amount than most other Texas cities. County Treasurer Waters had a policy of this kind on his residence, but not upon his other houses. The $2000 will more than cover the damage to his home.

CHAPTER X.

UNDER MARTIAL LAW.

HOW ORDER WAS RESTORED AND THE CITY WAS POLICED BY THE MILITIA.

Tuesday brought a substitute of State authority for the committee of safety. The adjutant general of Texas, Gen. Thomas Scurry, arrived to make a personal report upon the situation. To him the mayor and the citizens transferred all authority. The volunteers were disbanded. State troops were brought and camped on the wharves. Military discipline and etiquette went into effect. General Scurry appointed as his adjutant a newspaper man, Col. Hunt McCaleb, who had held a commission as lieutenant colonel in the Spanish-American war.

Thereafter men in uniform patrolled the city. A man without a military pass was in the wrong place. Bayonets bristled. Salutes were in order. "Halt!" rang out in the darkness if one ventured far after night. Those who were successful in getting into Galveston passed between the guards as they walked down the gangplank of the boat. Exodus was attended by the same presence of the military. Successive steps toward disposition of the bodies, clearing of streets and restoration of normal conditions were marked by

TEXAS CITY WHARF.

PIER 21.

general orders issued from headquarters. Galveston was a great camp, and was so governed until toward the close of the second week.

Another phase of the authority for the storm period was highly commendable. When two days after the storm a route into and out of Galveston by way of Texas City was opened a mob of people anxious to reach the scene besieged the train and practically took possession of it. At this juncture United States Marshal Grant of Texas appeared—another case of the right man in the right place. At the request of the railroad people and the citizens' committee, Marshal Grant took charge of the route between Galveston and Houston. The department at Washington honored a request for fifty special deputies. The marshal picked his men from among those he knew to be reliable. Thereafter nobody boarded a train unless he had such business in Galveston as warranted his presence there. Deputy marshals guarded cars and boats. They kept their eyes on the mail bags carrying currency and drafts and checks for large sums for relief. They did detective duty on the wharves of Galveston, scrutinizing everybody who landed from schooners or who came in by unusual routes. They were on duty continuously at Texas City. The surveillance was perfect. That there was need of it conditions at Houston showed. That city was full of dangerous people, crooks from other cities, the human vultures who were on the way to Galveston and would have gone there but for the impediments which Marshal Grant and General Scurry put in their way.

The system worked admirably. After the severe lesson which the volunteers administered in the first forty-eight hours lawlessness was not added to the other woes of Galveston.

The evolution of authority, law and order out of the confusion of the storm furnishes not the least interesting chapter. "Martial law" existed for twelve days. Just what that meant has not been fully explained. When the survivors pulled themselves together on Sunday morning they faced several emergencies, each of which called for courageous and wise leadership. Thousands of dead lay in the streets and among the ruins of buildings. In this semi-tropical climate quick disposition of bodies is indispensable for the health of the living. A city was without food. Survivors, with bleeding cuts and ugly bruises, were everywhere. A famine of water threatened. Not a dry building was in the whole of Galveston. Numerous as were the dead, there were more of the living than could be sheltered in the buildings still standing. But the men of Galveston faced still another problem. Out of the police force eight men were left, and they had exhausted themselves in life-saving work through the night. Every seaport has its lawless element. Galveston was no worse than any city of its class. Perhaps it was better than most. Certainly the record of the past fortnight speaks well. But the necessity of precaution against disorder was one of the first things to which the citizens gave consideration when they came together.

The chief of police, an old veteran, who had worked heroically to save lives, told of the condition of his force. Not many words were wasted. A committee of safety was formed by common consent of all present. J. H. Hawley, the general agent of the International & Great Northern Railroad, was made chairman of it, with the authority of the meeting to use his own judgment. His first act was to call for the Galveston Sharpshooters to re-enforce the eight policemen. Twenty-three riflemen were all who could answer "Here!" Out of Captain Rafferty's well-trained Battery O, First Artillery, 102 strong, only thirty-five had been able to reach the city at that time. They were without arms, and most of them without uniforms, being clad in such garments as they had been able to find. The regulars accepted temporary guard duty under the committee of safety. Then Chairman Hawley began to enroll a volunteer force from among the most responsible citizens he could find. Gun stores were entered and arms of every description were borrowed to equip the force. Into companies the 300 men were divided. They were lined up for instructions.

"You will remain on duty where you are assigned. You will protect the living and prevent depredations of property If you catch anyone looting houses or robbing the dead, you will shoot."

With these instructions the volunteer guards were marched to various localities and put on duty, until all parts of the city were under patrol.

This volunteer force remained on duty from Sunday until Tuesday. In the accounts sent out from Galveston covering the events of the two days were narrated instances of men being killed for looting. Some of the accounts entered into details. These stories grew after they left Galveston. When they had been repeated several times between the island and Houston they had been much magnified. The looters summarily dealt with increased in numbers like Falstaff's men in buckram.

Was there anybody shot in Galveston for robbing the living or the dead? The evidence would not convict in court. Circumstances and hearsay statements of reliable persons point to a limited number of cases in which summary justice was dealt. There are no official records. It was not intended that there should be any. Men in whom confidence was felt were put in charge of the companies of volunteer guards. They were told by the committee of safety that if anybody was caught in the act of looting no attempt need be made to take prisoners. If there was shooting the criminal should be buried where he was killed and no official report was wanted. On the best information obtainable it appears that between Sunday and Tuesday eight or ten persons fell before the weapons of the guards. After that it is doubtful if the lesson had to be repeated. The Galveston News' edition for Tuesday was a single small sheet of three columns. It contained this paragraph, which is believed to be well founded:

"The soldiers and police are instructed to shoot anyone caught looting or attempting to loot. The jails are full, and summary measures are necessary. The shooting of eight negroes was reported last evening. One soldier at guard mount reported that he had been forced to shoot five negroes. They were in the act of taking jewelry from a dead woman's body. The soldier ordered them to desist and placed them under arrest. One of the number whipped out a revolver and the soldier shot him. The others made for the soldier and he laid them out with four shots."

An illustration of the martial law under which the work has been done is General Order No. 9, issued by Brigadier-General Thomas Scurry, commanding the city forces, as follows:

"Guards, foremen of gangs and working parties or others acting under the authority of this department will use due diligence toward preventing any hardships on private individuals or impressing for service. The conditions, however, are so terrible, and it is so necessary that sanitary precautions be taken to preserve the lives and health of the people of this stricken city, that individual interests must give way to the general good. If it is found feasible to secure volunteers, general impressment will be avoided, but the medical fraternity being a unit in the opinion that further delay or procrastination will bring pestilence to finish the dire work of the hurricane, the interests of no individual, firm or corporation will for one instant be spared to secure volunteers for work."

Not twenty-five persons got into Galveston a day from the outside world. The lines were drawn tight, and pressing indeed must be the business which gave one safe passage.

The trains running to Texas City went down light. Nurses, physicians and soldiers, of course, came and went at will.

There were about 400 soldiers in Galveston doing guard and police duty. The camp on the wharf was rapidly put into shape, and the soldiers were soon comfortably housed. There were in Galveston five commands—the Dallas Rough Riders, Capt. Ormonde Paget, with forty-five men; the Houston Light Guards, Capt. George McCormick, with forty-five men; the Galveston Sharpshooters, Capt. A. Bunschell, with thirty-five men, and Battery D of Houston, Capt. G. A. Adams, with fifteen men, and Troop A, Houston Cavalry, commanded by Lieutenant Breedlove, with twenty men. Captain McCormick of the Houston Light Guards was acting major commanding the battalion. He said when he arrived that he expected the men would be needed in Galveston for several weeks yet. They had been doing splendid work, and he was more than satisfied with them.

On September 16 General Scurry had charge of the town, and it was really under martial law. Of course, there was some friction, because martial law means unusual restraint. But the friction, like the martial law, was a matter temporarily. It would be difficult to challenge the necessity of this measure. There were many defenseless women and children in the city, living in houses without locks and keys, and they had to be protected against prowlers of all kinds. How long such protection would be necessary no one knew at the start, but General Scurry could be depended upon to discharge the important obligations which he had assumed.

There were political factions who resented the idea of martial law, but this fact did not for a moment abate the necessity for it.

United States Marshal John Grant arrived on the fifteenth with twelve deputy marshals. He tendered his services to General Scurry, and they were accepted. He cancelled his political appointments in Ohio to render this service to Galveston. Speaking of the disaster, he said:

"It is the tragedy of the century, and is impossible of description. I have never seen anything like it before, and I hope I never shall again. As sorrowful as it is, however, I do not believe the people of Galveston will give way to despair. There is still a great future for this city, and those who survive must wisely utilize the present and build to the future. Such destruction is impossible of repetition, and all Texas will regret it if Galveston halts and refuses to improve the possibilities within her grasp. Of course, we who were not present cannot understand or appreciate how depressed such a stricken people must feel, but we are in position to add to our sympathies a fervent hope that in due time the people will decide to retrieve their fortunes and rebuild their homes. The horrible past—and thank God it is past—with its innumerable heartaches, is too awful to discuss. I would like to draw the attention of the people away from their misfortunes to the more hopeful future, which is no less hopeful because of the great sorrow which now overshadows all else, not only here, but throughout the country."

Men went to work clearing the streets of piles of timbers

and refuse. Men began to realize that the living must be cared for. It was the supreme duty. There was much work to be done, and it was being done. Women and children were hurried out of the city as rapidly as the limited facilities of transportation would permit. The authorities and committees were rational, and idleness was not permitted. There was an element with an abundance of vital energy who intended to save the town, and the town was saved. Burying the dead, feeding the destitute, cleaning the city and repairing wrecks of all characters assumed fair headway and was pushed as rapidly as men could be found to do the work. The great utilities of the city were repaired to a state of usefulness, men were in demand and workers came to engage in the duty of restoration. Life began to supersede death, and there was apparent everywhere a desire to save the city and rebuild it. Before one week had passed the listlessness of mourning people changed into a lively interest in life, and as this became so Galveston began to realize just what the world expected of her.

The situation during this period was well described in a dispatch by General McKibbin to the adjutant general in Washington:

"A complete organization for systematic work has been made; General Scurry, Governor Sayers' adjutant general, is its head. All other bodies are working under his orders. The city needs money and disinfectants. The surgeon general, through the medical associations throughout the coun-

WHAT IS LEFT OF THE RESIDENCE PORTION FROM THE URSULINE CONVENT TO THE BEACH—A DISTANCE OF SIX BLOCKS

try, could render great assistance by shipping disinfectants. There are plenty of doctors here."

There was plenty of work for ten times the force of laborers employed. The area which was untouched for more than a week embraced four and a-half miles of frontage on the beach and bay, and before it was cleared the bodies which lay rotting beneath the tangled timbers fell to pieces. Adjutant General Scurry, who was in supreme control, was unable on Saturday after the storm to pay all the laborers for their services. He ardently desired to do so, and tried to impress upon the committee the need for money.

The work done was remarkable, and was accomplished under the greatest difficulties. Track was laid along a right of way which was swept by the sea and washed into ravines, along a line bestrewn with dead bodies of men and animals. The men worked under a blazing sun, in water, and slush, and mud, in surroundings sickening to the senses and at first without adequate supply of food. The greatest difficulties in the way of securing material for the work, or rather in getting the material to the places where it is needed, were encountered. Notwithstanding these difficulties, the work went on day and night, and the structure that was to bring Galveston in real touch with the outside world was steadily pushed forward.

State Health Officer Blunt left on Sunday for Austin, where he made a report to the Governor concerning conditions in Galveston. This statement estimated the mortality at 8000 souls. Dr. Blunt was in Galveston several days, and made a thorough investigation of the results of the storm.

Horse cars were in operation in the business part of the city one week after the storm, and the electric lines and water service were partly resumed. The remaining parts of the city were not put in anything like its normal condition, of course, but order and system prevailed, and the people who were giving Galveston such noble assistance had good reason to be satisfied with what was accomplished in the face of such fearful odds. According to General Scurry, Mayor Jones and others, the progress of the work during the first week was more than satisfactory.

Tuesday the board of health began a systematic effort to obtain the names of the dead, so that the information could be used for legal purposes and for life-insurance settlements. Charles E. Doherty was stationed at the headquarters of the central relief committee to receive and file sworn statements in lieu of coroners' certificates. Persons who had left the city, but were in possession of information concerning the dead, were requested to send sworn statements to him.

Mayor Bradshear of Houston sent a corps of 105 volunteer street cleaners, who did splendid work under the direction of Matt Druman and Mr. Freeland. They were set to work in the downtown district, using a lot of coal drays, and by their help the business section was rapidly put in good condition.

The first street car was put in commission at 1 P. M. on Saturday, September 15, and began making trips from Market street and Twenty-first street to Broadway, and thence to Fortieth street. But three trips were made during the afternoon. The track was so covered with mud and grass that it was exceedingly difficult for the "hay-burner" motor to propel the car.

The car was in charge of Conductor R. H. Barrow and Driver R. F. Clarke and Mr. J. J. Stump of the street-car offices. The car bore the lucky number "66," and a mule known as "Lazy Lil" made the first trip. The little car slowly moved through the streets, and was hailed with cheers.

A few days later a few electric cars were fitted up and set to running, and by degrees the street-car service resumed something of its former efficiency.

The running of mule cars caused as much comment as the running of the first electric car in Galveston, and the turning on of the electric power in a small portion of the city, by which some of the streets and stores were lighted on Saturday night, gave the city a much more cheerful aspect than it has presented at any time during the week.

The day after the storm those who picked their way through the streets, climbing over huge piles of debris, said it would take many weeks to get the streets open so as to allow vehicles to pass. Houses were standing in the middle of the streets, telegraph poles with their entangling wires were in the way, fallen trees and wrecked matter of all kinds were in all the thoroughfares.

One week later a correspondent traversed a large part of

the city, and not only did he find almost every street north of the badly-wrecked portion along the beach passable, but many of the gutters had been opened and the water drained off. All of the business streets were open.

Two wires were working out of Galveston on Sunday, and while it was almost impossible to obtain any degree of promptness in the transmission of dispatches, still communication was open.

Street-lighting wires were repaired as rapidly as possible, and the flash of the electric light each night after Saturday threw weird shadows over the piles of wreckage. Incandescent globes were used, the arc lights having been all destroyed.

One detail of men was sent to the city water-works to help get the pumping machinery in order; another was sent to help the Gulf, Colorado & Santa Fe in its track and bridge work; another was loaned to the Brush Electric Light & Power Co. Other details were put to work cleaning up various streets.

On Sunday, September 16, the first after the disaster, there was for the first time in half a century no stated religious worship in Galveston. The churches were in ruins, and the day relief of workmen went to the clearing of the streets as usual at 6 A. M. Some of the clergymen got their followers together in small numbers and talked to them of the terrible visitation of Providence.

Mass was celebrated at St. Mary's Cathedral, and was largely attended. Father Kirwin preached an eloquent sermon, in which he spoke of the awful calamity that had befallen the people. After expressing sympathy with the afflicted and distressed he advised all to go to work in burying the dead—that was their first duty—and to bring the names of the widows and orphans to the church, and to rest assured of their being cared for.

With the beginning of the second week the restrictions of martial law were much relaxed, and the people of Galveston and strangers in the city were allowed to move more freely about the city.

CHAPTER XI.

GALVESTON REDIVIVUS.

DESPITE THE BOMBARDMENT OF THE WAVES, THERE WAS NO EXCAVATION OF THE ISLAND.

THE EROSION WAS INSIGNIFICANT.

THE STORM DEMONSTRATED THAT HOUSES COULD BE BUILT THERE WHICH WITHSTAND THE ELEMENTS' WORST FURY.

WONDERFUL PLUCK OF THE PEOPLE.

On Wednesday, September 12, conditions in Galveston were reported as being much improved. The water supply from the mainland was then working, and while it could not be sent through all the mains, there was enough for all needs in those portions of the city where life and property had been comparatively safe.

Similar reports were sent out in the press dispatches on Thursday, and one significant remark made by one of the correspondents on the spot was that the people were recovering from their grief and stupor. One press writer said:

"Thursday night's sleep made the people a new people. The difference in their look and deportment from that of the day before was observed by everyone. The streets were filled with them, when on the day before the streets were silent of all except those who had the horrible work of taking care of the dead on their shoulders. Now women could be seen talking to women. They met on the corners in the residence portions of the town and told their adventures. The men began to discuss the future. By 10 o'clock the town was up and buoyant. The effect of that one night's sleep was marvelous. There was no longer any talk of abandoning the city. Galveston should be greater than Galveston had ever been. That was on the lips of everyone.

"On Wednesday I would not have given $10 for the place. On Thursday I would have given more for a lot than I would have given before the deluge and storm. Why? Because the pluck of the people came out through that night of rest. Galveston should be greater than it had ever been. That is what they said. Galveston was safer than before by the island's weathering such a storm. That is what they said, too. They began to talk of their own pluck. We have stood much, but the world will say that we stood it well. If we can do as we have done in such a trial, what can we not do in the battle of life? Galveston shall be rebuilt. Galveston shall be the greatest of towns. Hurrah for Galveston! Thus they talked and went about their work of throwing up breastworks against disease by cleaning the town. Thousands of the people, negroes as well as whites, went about the work of burning the dead and cleaning away the debris. They asked nothing about wages, even those who had no property. They had begun the fight. It was evident that they intended to keep it up. The cold, calculating speculator would have had something to study over if he had seen these people as I saw them the day after their one night's rest. Well, there was nothing wild in their determination. The island has not a break in it.

"There is a story of millions of feet being torn from it and cast into the sea. This story may be true if applied to some part of the island which I did not visit. But where I went it is not true. There was no erosion. That was to be expected. Erosion would have come from a far less storm than this. I have seen a common 'rise' on the Ohio river carry away more dirt than this storm carried from Galveston Island into the Gulf. The people of the interior know where the old Beach Hotel stood. They know where the chimney of that house was built. They know how far it was from the beach. They will understand the work of erosion. I stated that the brick of that chimney is not in the water. The piling on which the hotel was built are in some places in the water. In fact, according to my observation, the erosion at this point has not been above 300 feet. I went to the east end of the town and to the west end of it. The destruction of the island is no greater anywhere that I saw than at the location of the hotel mentioned.

"For years and years people have said that when the right kind of storm came the island would sink under it or be washed away like a house of cards in a flood. It was supposed that the great currents which would rush across the

REFUGEE CAMP, FURNISHED BY THE U. S. GOVERNMENT.

ST. PATRICK'S CHURCH, THIRTY-FOURTH AND AVENUE K—HAD THE HIGHEST STEEPLE IN THE CITY.

33D ST. PIER.

VIEW FROM 14TH AND K.

island would dig bayous as deep as the bay. These would grow in width, and finally the great island would be cut into small ones, if it did not disappear beneath the waves. But what was the result of this greatest storm on record? Why, there is not, as far as I could hear, and I made inquiries, a single excavation made from the Gulf to the bay or the bay to the Gulf. The island stands there in all things, except in the matter of the erosion mentioned, as stable and firm as it has ever been since man knew it. That is enough. The foundation is there. Man can do most anything with a proper foundation.

"The only need now is stable and the right kind of houses. The old houses seem to have stood the shock better than the new ones. The reason of this is apparent. The old ones were built with an eye to storms. The new ones were built in book times. One young fellow told me that his house, the one in which he was born, had stood the storm of 1875 and every storm since that time without a quiver. 'And it would have stood this one had it not been for one thing,' he said. 'That thing was the outward flow of the tide when the storm was over. The water rushed back to the sea like a torrent. It fell over a foot and a-half in fifteen minutes, and as it went out it swept many a house from its foundations.' This flow, running like a torrent, swept across the island, and yet there was not left a single evidence in the way of excavations of its going.

"Attention was attracted to the house of Mr. J. H. Hawley, the brother of Congressman Hawley. He bought the property from an engineer who lived in Galveston some time about the flood of '96. He said he would build him a house which would stand. He placed the foundations of an iron fence two feet in the ground. This foundation was of brick. In this foundation he placed the railing of the iron fence running up three feet. At the top he placed fillagree brick work. His house was braced well and the timbers were heavy and well put together. The storm did not phase it. The fence acted as a barrier to timbers from the houses which had been destroyed. It kept away the battering rams with which the waves assaulted all places. When the night's horrors were at an end the house stood intact. Even the cistern, which was on piling, stood the test and was uninjured. Now the Galveston people begin to consider the question of whether much was not their own fault in that their structures were not of the kind that should have been built when storms were sure to come.

"It is just such things as this which give them hope. As I have said, I despaired of the town when I walked among the dead bodies and saw the destruction on every side. But like the rest I got over this depression. I caught the infection of the new life when it came. I know that I speak the truth when I say that the life in Galveston now is capable of upbuilding the town and building it better in every way than it ever was. Millions of dollars are invested in enterprises in the town. The men who have lost thousands, not to say millions, will not permit the rest to go without a struggle. The railroads running into the place and depending on the thirty feet of deep water, which is said now to exist in the channel, for export of the freight, will not agree to abandon the port, the only one of such depth for thousands of miles. Cotton factors in all the world, who look to this port for their supplies, will not abandon it. The monetary interest in the city of itself would save it even if the people were not so full of heart as they are. But above this, the poor people and the working classes have nowhere else to go. With many of them it is too late in life to begin it anew. It is too late for them to build up acquaintances again. They have lost their houses, but the lots on which the houses were located are there. Subscriptions to the amount of perhaps $2,000,000 have poured in for their relief. The well-to-do Galvestonian is determined that this relief shall go to those who are poor, that they may to some extent repair their fortunes. The rich themselves will build. In a month from now every man in the place will have all the labor he can perform. Every person will be busy. The work of upbuilding will in some measure rub out the recollection of the horrors of the storm The Huntington estate will continue its work. Bridges of the very first class will span the waters between the island and the mainland. If great corporations can risk their money, as they are determined to do, why shall not a poor man risk his labor to build another house on the lot he owned?

"Why, even behind the business and necessitous phases of the matter there rises a sentiment among the people. That sentiment is that we will show the world the stuff the Galveston people are made of. Galveston is all right. The storm could not kill her, though it wounded her to the death almost. There is pluck there. There is pride there. There is money there. And above all, there are recollections there for the Galvestonian, and he will not be downed by wind and wave."

On Thursday thousands of men were at work removing the debris and burning it. Stores were open for business and others were preparing to open by transferring their stock to the sidewalk to dry. Miles of street were lined with damaged goods drying under the fierce rays of a tropical sun. Every man, proprietor, clerk and porter, coatless and with trousers rolled up, was busy at cleaning up.

The spirit of the people of Galveston was there—one of hopefulness that the future city would rise greater and better than ever. Even the laborers and the workmen were imbued with this hope, and the thousands employed in burning the wreckage did their work with a will.

Fires were burning incessantly along the beach for two miles west of Tremont street, consuming wrecks of houses and the bodies of such victims as were still confined beneath. The poisoned air was becoming pure, except in the West End and along the beach and deep water front. Fire and disinfectants did wonders in those days.

"In thirty days you will not know it is the same city," said John Sealey, chairman of the finance committee. "We have sent over the State for 1000 carpenters, 1000 additional masons and artisans of all kinds. Galveston will rise greater and better than ever. This is what we desire to impress upon our good friends of the North and East.

"We are more than thankful to them for the substantial manner in which they have come to our aid, but much depends upon ourselves. It requires great trials to test the character of men. Never have I read or heard of a people responding with more pluck and more hopefulness than the people of Galveston.

"There is a great work to be done, and they will do it. Losses are forgotten, though the majority have lost everything that is dear to them. They now face the future with full confidence in it."

Equal confidence was shown by others among the men who control the finances and commerce of the city. The activity, energy and pluck displayed in the streets would have been marvelous anywhere, and it made Galveston over.

Most of the men who survived the Galveston storm and who determined to leave by this time had got away. It is estimated that about 5000 departed from the city soon after the storm, including women and children.

The resumption of business was actually forced upon Galveston. Six ocean steamships had anchored in the harbor since Tuesday awaiting grain cargoes.

Mr. Robertson, the grain inspector, said: "We will begin loading the steamers next week. All this wheat—2,500,000 bushels—will be saved if we have no rain. Colonel Polk, general manager of the Santa Fe, tells me he will have trains in Galveston next week. We never say die here. Look up and down these streets and you will see an example of the greatest hustling ever known."

Preparations for rebuilding were begun in the business part of the city, the railways and water front being rapidly cleared of ruins. The telephone and telegraph companies were rushing things. The Western Union had five wires strung to its downtown offices. The Postal had some up by Friday evening, and full telegraph service was established by the close of the week. Business on the floor of the Cotton Exchange was established in three weeks. The building was badly damaged and partly unroofed.

So much progress had been made toward the rehabilitation of Galveston and so harmoniously were the various forces working, that General McKibben, who was ordered to Galveston with his staff to assist the authorities as soon as the storm disaster befell Galveston, decided that his presence was no longer necessary, and he made arrangements to leave for Houston. After having largely assisted in the restoration of local confidence, the withdrawal of General McKibben was taken to mean that little was to be done but to take care of the distressed until normal business conditions had been resumed. In this connection the information was made public through the local representatives of the federal authorities that the War Department would undertake as soon as possible the restoration of its property at this point.

The idea that the status of the city would be changed as a result of the disaster found no local adherents. The various railroads entering the city determined to assist the citizens of Galveston to the full extent of their ability in rebuilding the city. Col. L. J. Polk of the Santa Fe received a very enthusiastic and encouraging message from the headquarters of his road, declaring confidence in Galveston, urging the business community to proceed at once with the work of reconstruction and promising every help in their power. Speaking upon this point, he said:

"The railroad interests have decided to combine their forces in order to rebuild as quickly as possible a bridge from Virginia Point to Galveston. A large number of men will go to work in the morning with this end in view. You may say to the country that in six days a bridge will have been built and trains running over it. I have had a consultation with the wharf interests, and they have promised us that they will be prepared to handle ingoing and outgoing shipments by the time the bridge is finished. The bridge we shall build will be of substantial but temporary character. We shall subsequently replace it with a more enduring structure. There is no reason why Galveston ought not commercially to resume normal conditions in ten days."

It was noticeable that the people no longer dwelt upon the number of those that were gone. A soldier standing guard at a place on the beach where the fires were burning thickly was asked if the workers were still finding bodies. "Yes," he replied, "a good many." That was all.

Three days before the same soldier would have gone into particulars. The commander of one of the squads came into headquarters to report. He had nothing to say about bodies, but wanted to tell that a trunk in fairly good condition with valuable contents had been taken out of one heap, and that the owner might be found through marks of identification which he had noted.

So it went. The thought was of the living rather than of the dead. Passing along Tremont street and looking up and down the crossing streets one saw hundreds of wagons and carts being loaded high with the fragments of building materials. As quick as this refuse could be taken up it was hauled to vacant spaces and added to the bonfires, which burned continuously.

Galveston went through a kind of purification by fire. Full of confidence, and even optimistic, were the expressions of the men who had taken the lead in this crisis. Colonel Lowe of the Galveston News said: "In two years this town will be rebuilt upon a scale which we would not have attained so quickly without this devastation. I take it for granted that when the Southern Pacific management says to its representatives, as it has said, 'Build a bridge ten feet higher than the old one, and put on double force to do it,' our future was assured. We shall go forward and re-create the city. We shall have some restrictions as to rebuilding lines, especially on the beach side, where the greatest losses were sustained. The ramshackle way in which some construction has been done in the past will be of the past."

It must not be inferred from this that as the days passed Galveston found losses of life and property to have been exaggerated. It was thought sufficient on Sunday to place the

NEAR PIER 16.

dead at 1000. Nearly a week passed and the conservative—it might be said the semi-official—statements swelled the number to not less than 5000. Indeed, Mr. McVittie expressed the fear that the total might be 6000. That is out of a population of 39,000.

It took on the part of these public-spirited men a good deal of boldness to lay down the law that the support tendered by the country must be earned in order to demand it. Some people in Galveston felt that they had claims upon the charity, and rather resented the decision which compelled them to render an equivalent in labor. But the majority backed the leaders in their stand. Before two days had passed the whole community was at work cheerfully.

A tour through the city, up one street and down another, showed the greatest activity. Thousands of men were dragging the ruins into great heaps and applying the torch. Occasionally they came upon the remains of human beings and hastily added them to the blazing heaps. It is notable that much less was then said about the dead than during the early days of the week. The minds of the people who survived passed from that phase of the calamity.

In striking contrast to the first feeling of despair after the storm was the hope and determination which the people of the stricken city were now showing. Work was pushed under a systematic plan of operations that rapidly brought order out of chaos. The search, burial and cremation of the unfortunate victims within the corporation limits of the city was rapidly carried on by large forces in organized squads under military direction.

Dry goods stores and clothing houses resembled great laundries, and every available space was occupied with goods hung out to dry. Fortunately, the weather was clear, hot and dry for this purpose. Those merchants whose stocks were but slightly damaged did a rushing business, and so did the restaurants, whose stocks were very limited. Fresh meat was difficult to obtain. Extortion was, however, rare, although the supply of food at hotels and restaurants was meager. This was overcome in a few days, when all the railways terminating had united upon one bridge and went to work night and day with a large force reconstructing it, while the tracks were being restored on the island and mainland. This gave rail communication by the middle of the second week, and the situation was then much relieved.

Governor Sayers expressed faith in the future of Galveston, and said the city would be rebuilt. He wanted the people of Galveston at once to turn their attention to the rehabilitation of their property, and leave to the Governor, General Scurry and the State authorities the work of sanitation. In other words, the State would relieve Galveston of the important work of sanitation, and would leave the citizens free to restore their homes and their places of business.

With reference to the rehabilitation of Galveston, Representative Hawley said in Washington:

"The city of Galveston is the unrivaled port of all that territory lying west of it and as far north as Omaha, Neb. It serves millions of people as their gateway to the sea, and has facilities to accommodate millions of tonnage for carrying the products of this vast territory to every part of the world. The recent storm which swept the Gulf was without a parallel in our history, and during the few days that its destructive force interfered with the facilities for shipping in Galveston there was congested an enormous amount of business in the State of Texas and north and west of it.

"Immediately after the storm Colonel Riche, the alert and able United States engineer in charge of the harbor works, took soundings to determine the effects of the storm. He at once reported that the port was free and unobstructed, with a safe depth of twenty-seven feet.

"While mourning the dead, all industrial life of Galveston was promptly in motion to re-establish and maintain unbroken all the avenues of commerce, domestic and foreign, and the unfaltering courage, hope and purpose of Galveston's citizenship shone out in splendid contrast with the ruins which laid so heavily on many homes about them. The established wharves were found also to be injured only in the superstructure, which contributed largely to the quick re-establishment of trade relations.

"With every energizing force at work, Galveston will rise and resume her sway. Among the prime factors that will contribute largely to this restoration of former conditions on a more substantial basis is the universal sympathy and co-operation of generous men everywhere in modifying the awful effects of our fell disaster. Another most essential element will be the resumption of work by the government in the restoration of all its aids to commerce throughout the harbor and of its fortifications, so necessary to its defense.

"It is not Galveston as a city alone that speaks for these conditions," he said, in conclusion, "but for that immense community of interests which lie behind it and use it in the disposition of all their products through the carriers that come and go to and from the markets of the world."

But no greater evidence of the faith of the people of Galveston in the resurrection of their fallen city can be found than the following letter written by the proprietor of the Galveston News to the New York Herald:

"Galveston, Texas, Sept. 14.—New York Herald, New York: Replying to your telegram of the 12th instant, the management of The News appreciates your sympathy and kindly offer to extend the Herald's news service to us for six months gratis. Appreciating the kindly tender of gratuitous service, we are glad to say that this will not be necessary. The offer, however, is none the less appreciated on that account.

"You ask The News what is our estimate of Galveston's future, and what the prospects are for building up the city. Briefly stated, The News believes that inside of two years there will exist upon the island of Galveston a city three times greater than the one which has just been partially destroyed. The devastation has been great, the loss of life terrible, but there is a hopefulness at the very time this answer is being penned you that is surprising to those who witness it. That is not a practical answer to your inquiries, however.

"The practical feature is this: The Southern Pacific Railway Company has ordered a steel bridge built across the bay ten feet higher than the trestle-work on the late bridges. The Southern Pacific has orderd a doubling up of forces to continue and improve their wharves, and with this note of encouragement from the great enterprise upon which so much depends the whole situation is cleared up. Our wharves will be re-shedded, the sanitary condition of the city perfected, streets will be laid with material superior to that destroyed, new vigor and life will enter the community with the work of construction, and the products of the twenty-one States and Territories contiguous will pour through the port of Galveston. We have now through the action of this storm, with all its devastation, thirty feet of water on the bar, making this port the equal, if not the superior, of all others on the American seaboard. The island has stood the wreck of the greatest storm convulsion known to the history of any latitude. There is no longer question of the stability of the island's foundation. If a wind velocity of 120 miles an hour and a water volume of fifteen feet in some places upon the island did not have the effect of washing it away, then there is no wash to it. Galveston Island is still here and here to stay, and will be made in a short time one of the most beautiful and progressive cities in the Southwest.

"This may be esteemed simply a hopeful view, but the conditions existing warrant the acceptance of the view to the fullest extent.

"The News will not deal with what is needed from a generous public to the thousands of suffering people now left with us. The dead are at rest. There are 20,000 homeless people here whose necessities at this time are great indeed. Assistance is needed for them in the immediate present. The great works of material and industrial energy will take care of themselves by the attractions here presented for the profitable employment of capital. We were dazed for a day or two, but there is no gloom here now as to the future. Business has already been resumed.

"THE GALVESTON NEWS."

The New York Commercial's comment upon the situation in Galveston at this time may be taken as representative of what men of judgment outside thought of abandoning the storm-swept island:

"The cry of despair that comes up from Galveston is not of the survivors among her own people. It cannot be that they regard their city as beyond recovery from the blow that has temporarily paralyzed her. Galveston has suffered more than once before from the fury of the elements, but never so severely as now. Still, her people are not of the sort that are ready to lie down and die in the face of any disaster short of the absolute destruction of the city and the site itself. Barring the exposed position—a natural disadvantage by no means insurmountable by modern engineering—there is every reason why the great commercial port of Texas should rise from he ruins and speedily enter upon a new era of growth and progress and vast usefulness in the business world.

"Here is a population of 45,000, a tithe of which may have been wiped out by the tornado, but the brains, the energy, the ambition, the hopes of Galveston's people remain. They need only a chance to recuperate, and recuperate they must. This chance is within the gift of the business men of New York and of the entire country. Galveston needs something more than the succor that will alleviate the physical sufferings of her people. Her merchants and captains of industry need money, credit, the helping hand that pulls the weak over rough places, and the encouragement that stimulates to self-effort and final success. All this they ought to have, and will have, the Commercial firmly believes, from the business men of the United States.

"By no city of her size in the world has Galveston been surpassed in the energy and the enterprise of her people or in the honorable records of her business men. They should not be sacrificed at this time. It would be manifestly unjust to exact the 'pound of flesh' when they are suffering and helpless, as at present. Furthermore, it would be not only just and charitable to extend credits and give every other form of assistance possible, but it would be the part of wisdom, too, and good policy. It would be bread upon the waters which would be returned some day tenfold to the merchants and financiers of the country.

" 'Without this help,' says our Houston representative

'few firms can survive. With it, many can recover.' The duty of the business community is plain.

" 'Galveston,' says the chairman of the Southern Pacific Company's directorate, 'is geographically the port for our company. We shall begin to rebuild at once. I do not believe the reports that the city generally will not be rebuilt, for the reason that Galveston is wonderfully rich in resources and its situation makes it the natural commercial outlet for all that section of the country. The disaster will not kill Galveston as a shipping port.'

"This is the position of a rich and resourceful corporation that needs no direct help. The same spirit unquestionably animates the business men of Galveston, less fortunately circumstanced. The vast wealth and the great heart of the country must supply the deficiency."

LARGE HOUSE TO LEFT MOVED THIRTY YARDS WITHOUT INJURY—NINETEENTH AND N½.

TREMONT AND AVENUE L, LOOKING SOUTH.

CHAPTER XII.

STORM SCIENTIFICALLY CONSIDERED.

IT AROSE IN THE WEST INDIES AND CAME ACROSS TO FLORIDA AND TEXAS.

HOW THE OPPOSING WINDS AND TIDES WARRED OVER GALVESTON ISLAND—A PART OF THE CITY PROTECTED BY THE INCREASING PILE OF DEBRIS.

The West Indian storm, which developed into a hurricane after entering the Gulf of Mexico, and which fell with destructive force upon the coast of Texas, was similar in its scientific aspects to many others which have preceded it.

The United States Weather Bureau was informed that the first sign of the disturbance was noticed August 30 near the Windward Islands. August 31 it was still in the same neighborhood. The storm did not develop any hurricane features during its slow passage through the Caribbean sea and across Cuba, but was accompanied by tremendous rains. During the first twelve hours of September 3, in Santiago, Cuba, 10.50 inches of rain fell, and 2.80 inches fell in the next twelve hours. September 4 the rainfall during twelve hours in Santiago was 4.44 inches, or a total fall in thirty-six hours of 17.20 inches. There were some high winds in Cuba the night of September 4.

By the morning of the 6th the storm center was a short distance northwest of Key West, Fla., and the high winds had commenced over Southern Florida, forty-eight miles an hour from the east being reported from Jupiter and forty miles from the northeast from Key West. At this time it became a question to the Weather Bureau officials as to whether the storm would recurve and pass up along the Atlantic coast, a most natural presumption, judging from the barometric conditions over the eastern portion of the United States, or whether it would continue northwesterly over the Gulf of Mexico.

Advisory messages were sent as early as September 1 to Key West and the Bahama Islands, giving warning of the approach of the storm and advising caution to all shipping. These warnings were supplemented by others on the 2d, 3d and 4th, giving more detailed information, and were gradually extended along the Gulf coast as far as Galveston and the Atlantic coast to Norfolk.

On the afternoon of the 4th the first storm warnings were issued to all ports in Florida from Cedar Keys to Jupiter. On the 5th they were extended to Hatteras, and advisory messages issued along the coast as far as Boston. Hurricane warnings were also ordered displayed on the night of the 5th from Cedar Keys to Savannah. On the 5th storm warnings were also ordered displayed on the Gulf coast from Pensacola, Fla., to Port Eads, La.

During the 6th barometric conditions over the eastern portion of the United States so far changed as to prevent the movement of the storm along the Atlantic coast, and it, therefore, continued northwest over the Gulf of Mexico. On the morning of the 7th it was apparently central south of the Louisiana coast, about longitude 89, latitude 28. At this time storm signals were ordered up on the north Texas coast, and during the day were extended along the entire coast. On the morning of the 8th the storm was nearing the Texas coast, and was apparently central at about latitude 28, longitude 94.

The last report received from Galveston, dated 3.40 P. M., September 8, showed a barometric pressure of 29.22 inches, with a wind of forty-two miles an hour northeast, indicating that the center of the storm was quite close to that city. At this time the heavy sea from the southeast was constantly rising and already covered the streets of about half the city. After that nothing was heard direct from Galveston up to Tuesday morning. No reports were received from Southern Texas, but the barometer at Fort Worth gave some indications that the storm was passing into the southern portion of the State. An observation taken at San Antonio at 11 o'clock, but not received until 5.30, indicated that the center of the storm had passed a short distance east of that place, and had then turned to the northward.

The next information received Monday morning was that it had reached Oklahoma, as stated.

Within the radius of the effects of the disturbance high

winds prevailed. September 6 the wind was thirty-six miles an hour at Charleston, S. C., from the east, while the same night in Port Eads, La., it blew fifty-six miles an hour from the northeast. At New Orleans Saturday the wind reached a velocity of forty-eight miles, and at Fort Worth Sunday of fifty-two miles.

The representative of the Associated Press at Houston on September 10 sent to Northern papers the following description of the storm:

"Tonight the city of Galveston is wrapped in sackcloth and ashes. She sits beside her unnumbered dead and refuses to be comforted. Her sorrow and suffering are beyond description. Her grief is unspeakable. Friday and Saturday she was happy, prosperous and buoyant, and with a bright and prosperous season opening up auspiciously. Last night she was stricken down and crushed by a misfortune that seldom befalls any community, and in her inexpressible anguish appeals for help to bury her beloved dead, feed her stricken and afford temporary relief for those who almost in the twinkling of an eye lost home, loved ones and the savings of a lifetime. Tonight the city is dark, desolate and dreary. A pall has fallen over the living. It is pitiful and pathetic beyond expression.

"The cyclone that produced such an appalling disaster was predicted by the United States Weather Bureau to strike Galveston Friday night, and created much apprehension, but the night passed without the prediction being verified. The conditions, however, were ominous. The danger signal was displayed on the flagstaff of the Weather Bureau. Shipping was warned. The southeastern sky was somber, the Gulf beat high upon the beach with that dismal, thunderous roar that presaged trouble, and it had that ominous stillness that betokens a storm. From out the north in the middle watches of the night the wind began to come in spiteful puffs, fitful at first, but increasing in volume as the day dawned.

"By 10 A. M. Saturday it was almost a gale; at noon it had increased in velocity, and was driving the rain, whipping the pools and rattlings things up in a lively and ominous manner, yet no serious apprehension was felt by residents remote from the encroachments of the Gulf section of the beach. As stupendous waves began to send their waters far inland the latter began a hasty exit to more secure places in the city.

"Two gigantic forces were at work. The Gulf drove the waves with irresistible force high upon the beach, and the gale from the northeast pitched the waters against and over the wharf abutments, shaking the sewers and flooding the city from that quarter. The hapless people were caught between the two incoming floods, while the wind shrieked, howled and rapidly increased in velocity.

"Business suddenly came to a standstill. Car traffic was impossible, and all those that had a home and could reach it either by conveyance or otherwise hastily left their places of business and offered fabulous prices for any kind of a vehicle that would carry them to their loved ones.

"Railroad communication was cut off shortly after noon, the tracks being washed up. Wire facilities completely failed at 3 o'clock, and Galveston was isolated from the world.

"The wind momentarily increased in velocity, while the waters rapidly rose, and the night drew on with apprehension depicted in the faces of all. Already hundreds were bravely struggling with their families against wave and wind for places of refuge. The public school buildings, the courthouse, the hotels, and, in fact, any place that offered an apparent refuge became crowded to their utmost. Darkness settled on the city like a pall, while the wind shrieked with frightful velocity, and the rain fell in torrents.

"At 6.28 o'clock P. M., just before the anemometer blew away, the wind reached the frightful velocity of 100 miles an hour. Buildings that had hitherto withstood the gale then collapsed, carrying death and destruction to hundreds. Roofs whistled through the air, windows were driven in with a crash or shattered by flying slate, and telegraph, telephone and electric-light poles, with their mass of wires, were snapped off like pipestems.

"What velocity the wind attained after the anemometer blew off is purely a matter of speculation. The heavy detonation of falling buildings and piercing cries for help that broke through the rush and roar of the elements, and the picture of dead bodies floating along the streets, made it a night to those who safely passed through it that will never be obliterated from their minds. The lowest point touched by the barometer in the correspondent's office, which was filled by frightened men and women, was 28.04½. That was at about 7.30 P. M. The barometer then began to rise slowly, and by 10 P. M. it had reached 28.09. The wind gradually subsided, and by midnight the storm had passed. The water, which had reached a depth of eight feet on Strand street, at 10 P. M. began to ebb. It ran out very rapidly, and by 5 A. M. the crown of the street was free of water.

"Thus passed out the most destructive storm that ever devastated the coast of Texas."

A storm bulletin was issued on Monday, September 10, by Willis L. Moore, chief of the Weather Bureau at Washington, giving the origin of the West Indian hurricane and its extent. The bulletin follows:

"Several times during the present summer there have been indications over the Lesser Antilles of the possible development of one of those tropical storms, which are the most dreaded characteristics of the locality and season, but it was not until the first day of the present month that the particular storm under discussion assumed definite formation and progressive movement. It is true that for two days previous falling pressure and some abnormal movement of the cirrus clouds had given slight premonitory signs of the approaching disturbance, but not more than on many previous occasions, and two days passed before the assurance came that a tropical storm was really in progress.

"On the morning of September 1 the storm was central south of the island of San Domingo in about latitude 15 degrees north and longiture 70 degrees west. It moved slowly northwestward, and by the morning of the 4th it was apparently central south of the middle Cuban coast in about latitude 22 degrees north and longitude 81 degrees west. Previous to this time the pressure had been falling steadily but slowly as the center of disturbance advanced, but there were simply torrential rains, without destructive winds.

"During the 4th, however, the direction changed to a more northerly course, the pressure began to increase more rapidly, and the winds to increase in force, and by the morning of the 5th the center had passed over Western Cuba to the channel between Havana and Key West, causing high winds over Western Cuba and extreme Southern Florida.

"At this time it became necessary to determine, if possible, the future course of the storm, as its center was now near the place where a recurving to the north-northeastward was probable. Decreasing pressure over the interior of the country east of the Mississippi river indicated that a path would be opened northward along the Atlantic coast, and warnings and advisory messages to that effect were accordingly issued. At the same time the barometric conditions over the Northwest were such that, with a rapid eastward movement of the high area then covering that section, the tropical storm would be forced to continue in the direction which it then had and proceed over the Gulf of Mexico. To guard against every possible contingency, advisory messages were also sent to all middle and west Gulf ports, giving full in-

VIEW FROM 15TH AND K.

formation concerning the movement of the storm center and its possible extension to their territory.

"On the morning of the 6th the center was slightly north of Key West, and northeast gales were general over Southern Florida, Jupiter reporting a velocity of forty-eight miles per hour and Key West one of forty miles. By the evening of the 6th the storm was near the Florida coast, a short distance south of Tampa. Then came a sharp turn to the west, and on the morning of the 7th it was in the Gulf of Mexico, apparently about 200 miles south of the Mississippi coast.

"Northerly gales were holding full sway over the middle Gulf coast, Port Eads, La., reporting a maximum velocity of fifty-six miles an hour from the northeast. The storm warnings which had been displayed on the previous day as far west as New Orleans were now extended along the Texas coast, and on the morning of the 8th the storm had nearly reached that locality. The coast of Texas was reached by the storm during that afternoon. At 3.40 P. M. Galveston reported a barometer reading of 29.22 inches, a wind velocity of forty-four miles an hour from the northeast, unprecedentedly heavy sea swell and high tide from the southeast, and about one-half the city streets under water. After that time no communications were received from any Southeastern Texas points for several days, except from the manager of the Western Union Telegraph Co. at Houston, Texas, who, in response to a telegraphic request from the chief of the Weather Bureau, stated in dispatches sent on the 9th that no definite information could be obtained from Galveston, but that the losses of life and property were most appalling.

"After reaching the coast the storm's center once more turned to the northward and continued through the State of Texas during Sunday, the 9th, with steadily decreasing intensity, although it caused high winds, which were a source of great danger to many growing crops. Monday morning the storm reached Oklahoma, but its destructive character was gone, and it was now principally engaged in causing general rains in its vicinity."

Monday night the forecast sheet issued by the Weather Bureau stated that the storm had moved from Oklahoma City north to central Kansas, where its intensity was greatly diminished. Heavy rains occurred in Kansas and in the borders of adjoining States.

From this point the storm proceeded northeastward to the lake region, and thence down the St. Lawrence and out to sea. During its passage after leaving the State of Texas it decreased somewhat in intensity, though it was marked by tornadoes in Illinois, one visiting Beardstown, where considerable damage was done. Shipping was lost on the lakes and some injury was done to crops and farm buildings along the southern shore of Lake Erie. The passage of the cyclone through the United States was not an unmixed evil, as it brought to an end one of the severest and longest-continued heated periods that ever visited the country.

A writer in the Galveston News, in describing the scientific aspects of the storm there, wrote as follows:

"I have attempted to figure out, as men would say, how this storm that bore down upon the doomed town on Saturday and subsided on Sunday morning last could have wrought the catastrophe it builded with demoniac hands. I have talked with dozens of men, and women and children, too, who came out of the maelstrom, and I have been amazed at the coolness and detail with which they have, one and all, related their experiences. No two have had the same experience. The man who impressed me most is Captain Stirling of the schooner of that name. He was raised on Galveston Island, and has followed the sea for twenty years or more. He is not an educated man as books go, but he has passed his life in the comradeship of wind and wave, and is as staunch a seaman as ever stretched canvas or reefed a sail. I would rather risk him than the man of books.

"The storm was inaugurated by a strong north wind, veering to northeast. The lowest tide in the memory of living observers obtained in Trinity bay or Double bayou, forty miles north of Galveston Island. All this tremendous volume of water was driven by the impact of the hurricane into Galveston bay and from thence churned diagonally across the Island City. Meanwhile there was another storm gathering to the south and east and racing down from the keys onto the Texas coast and lashing the waters of the Gulf forward to complete the devastation of Galveston. This was the course of the storm that swept the drowned and maimed and tons of wreckage from Galveston to the mainland, nine miles across the bay, and which spent its fury in the destruction of villages, farmhouses, farms and lives for miles along the Texas coast line.

"Galveston Island sets approximately to the points of the compass east and west with the bay, nursing in its tranquillity the menace of ruin to the island, whose side it laves with coy ripples, touching it on the north and the Gulf, with its eternal roar in which there is the suppressed growl of a tethered fury, lapping it on the south. This north wind from 12 noon on Saturday hammered away at the doomed island until Saturday evening at 7 o'clock, when it veered to the northeast, from which quarter till 9.20 it wrought out its havoc. Then succeeded the storm from the southeast, which blew to a speed of eighty-four miles an hour, when it smashed the government apparatus and left the record with the Storm King, who tells no mortal man of how fast he can travel. It was from that moment forward, until Galveston was reduced to ruins and not a house of all its thousands left unscarred by the sweep of the hurricane or the drive and tug of tide, as if some demoniac devil had bade Typhon to rend his soul in one supreme burst of infernal wrath upon that island strip where stood proud and confident Galveston. And Typhon rose and met the summons as mightily as he has manifested his strength upon any spot this old earth has known.

"It is not at all remarkable that of all the statements in regard to the details of this storm no two persons can be found who agree on the direction of the wind and the currents. All agree that the most terrible blows which the town received came from the point of the compass which may be spoken of as between northeast and east. There are those who declare that at first the wind was almost from the north. Then it veered till it was almost east, and then settled down to its herculean efforts from a point between the two; and yet there are others who say that it came from all directions at different times, and prove it by the loss of windows in their houses. These waves came in from the Gulf. They filled the bay. The water chased across the island, met the waves, and then it seems there was a battle between the two elements, for the currents ran criss-cross. They went down one street, up another street and across lots. They seized a house here and placed it there. Men were carried to sea. Men were carried down the island. Men were carried across the bay by it. No chart can be even dreamed of their peculiarities. The wind lashed the water and it fled. That was all there was in it, and it fled in every direction, carrying on its bosom a shrieking people. It carried, too, houses whole, houses in halves, houses in kindling wood. The winds dipped and seized the debris and hurled it on. The air was filled with missiles of every kind. The water held them and threw them from wave to wave. The winds grasped them as they were thrown and hurled them further. Stoves, bathtubs, sewing machines, slates from roofs—these were as light in the hand of the two giants, wind and water, now in their fury, as the common match would be in the hand of the strong man.

"From the northeast it is generally conceded the storm came. Galveston Island runs nearly east and west. So it will be seen that it had a clean sweep from end to end of it. The streets are numbered across the island. They are lettered as they run with the island, east and west. For instance, the street running east and west nearest the bay is A street. Then there is B, and so on toward the Gulf. P and Q streets may be said to be two-thirds across the island; that is to say, they are three-quarters of a mile from the bay and one-quarter of a mile from the Gulf. This is not an accurate statement, and is only given to illustrate. Between Q street and the Gulf were hundreds and hundreds of houses. While many were fine mansions, the great majority of them were the houses of the poor. Coming down the island from the east, the storm struck these habitations.

"It was in this area, east and west, from one end of the town to the other, it did its worst. The large houses were overthrown. Where they fell they were hammered into shapeless masses. The small ones were taken up. A man can take two eggs and mash them against each other. The waters took the remnants and pushed them forward. One street of buildings would go down. That would be next to the Gulf. The timbers were hurled against another street. It would go down. The debris of the two would attack the third. The three would attack the fourth, and thus on till Q street was reached. Here the mass lodged. It is said by some, though I know nothing of it, that about it is the backbone or high part of the island. The great mass of matter became heavy. It must have dragged upon the ground, as the water here could have only been five to seven feet deep. But this would not have stopped it had the last street to be assaulted, Q street, or Q½ street, not interposed. The houses here were rather large and strong. This battering ram made by the winds and worked both by the winds and the water, met with resistance from the houses and was impeded by its own weight, which dragged it on the bottom. Its efforts at destruction became more and more feeble. The houses stood, though wrecked. The debris climbed to the very eaves. But the more that came the heavier the mass became. And lo! the very assailant became the defender, for, piling higher and higher by the addition of houses lately splintered, by the addition of everything from a piano to a child's whistle, there was a wall built against the great waves which rolled in from the Gulf, and thereby the territory lying between the bulwark and the bay was protected to some extent. True, the casual observer will think, as he looks even up and down the main streets of the town, that very little protection was given. But few lives were lost, in comparison, in this district, and while the stores were flooded and houses toppled over by the winds and undermined by the water, yet that bulwark, made of dead people and all they had struggled for and owned in this life, kept back the savage waves from the Gulf and saved the rest of the town. Looking at this wall, from which come the odors of decomposition, climbing it, as this correspondent has done, he is sure in his mind that if it had not been formed not as many people of Galveston Island would have escaped as on that day when Pompeii was shut out from the eyes of the world by the veil of ashes."

In this storm the usual conditions were reversed, whereas in wrecks by wind or water first reports greatly magnify the loss of life, in the present case it seems that the estimate increased rather than diminished as each day passed. While the total will never be known, it will be far above the early estimates.

J. M. Cline, local forecast official at Galveston, made an official report to the Weather Bureau on the Galveston cyclone. He reported that on September 8 the city was visited by a hurricane. The lowest reading of mercurial barometer was 28.53, the lowest ever recorded in this country. The readings were taken by John D. Blagden to check the barograph. There was a total fall of one inch of rain in eight and one-half hours.

The usual premonitory signs heralding the approach of a hurricane were absent. Brick-dust sky was not observed. At 5 A. M. on the 8th he noted the heavy swell of the Gulf, and telegraphed Washington that the lower portions of the city were being flooded and that such high water, with opposing winds, had never been noted before. The water mounted steadily in the face of the wind. Dense clouds and rain prevailed during the afternoon of the 8th, with the wind steadily increasing, until it reached a storm velocity about 5 A. M. An hour later it was blowing eighty-four miles an hour for a four-minute stretch, and at 8 o'clock was blowing 100 miles an hour. The aerometer blew away, and the

CITY WATER WORKS AND ELECTRIC PLANT—NEARLY TOTALLY DESTROYED.

velocity of the wind was later estimated at 120 miles an hour. It then shifted from northeast to southeast and blew harder than ever.

Mr. Cline says Mr. Blagden looked after the instruments heroically until they were all blown away. The storm warnings were timely and received wide distribution, not only in Galveston, but all along the coast. Giving the high tide and storm signals kept one man busy at the telephone on the 8th. People were warned that the wind would go by east to south and the worst was yet to come.

As a result thousands moved away from the beach section to the center of the city and thus were saved. Fifty persons took refuge in Cline's house, and all but eighteen were killed, including his wife. The water rose steadily from 3 P. M. to 7.30 P. M., when there was a rise of four feet in four seconds. The total rise in the tide was twenty feet. Cline's house was very substantial, but was finally wrecked. He floated about for three hours, but was blown into shore and saved. Accompanying the report is a carefully-prepared map of the city, showing 3636 houses destroyed.

The property loss is estimated by Mr. Cline at $30,000,000, and the deaths at over 6000.

In commenting upon the storm a few days after its passage the New Orleans Times-Democrat said:

"We have had any number of these storms, and nearly all have been most destructive to life and property. Bred in the West Indies, they have swept up from the coast of Cuba, sometimes veering to the left and ravaging the Gulf coast from Texas to Florida, and again turning eastward and sweeping along the Atlantic coast as far north as Boston, then losing themselves somewhere in the Atlantic between America and Europe.

"The history of Louisiana is filled with the disasters of these hurricanes. We of this section can recall the Last Island horror, the Indianola storm, the Sabine Pass hurricane, the Johnson bayou disaster, and, last and worst, the Cheniere Caminada storm of seven years ago, and the Atlantic coast has been nearly as unfortunate.

"The year of the Cheniere Caminada storm, 1893, has the worst record. There were three violent and distinctive hurricanes that year—one on the Gulf and two on the South Atlantic coast, all of them causing great loss of life. The Atlantic storms ravaged Florida, Georgia and South Carolina, particularly the sea islands of the two latter States, and the list of dead ran far up in the hundreds.

"Since then the Gulf coast has escaped any disaster, but the West Indies and the South Atlantic have not been so fortunate. It was not so many months ago that one of the West Indian hurricanes played havoc in Porto Rico, Martinique, Guadeloupe and St. Lucia. In Porto Rico the losses were so great and the crops so badly injured that the United States were compelled to establish relief stations there, and the island is today suffering from the effects of that storm.

"Of this storm we had, as usual, ample notice. The Weather Bureau hung out storm signals, and as a consequence no vessels have put to sea of late. It was impossible to do more than this, but the people were put on their guard, and waited to see at what point the storm would strike, for, like lightning, it seldom strikes twice in the same place.

"It would seem that New Orleans is storm-proof—that is, sufficiently far from the Gulf and protected from hurricanes to escape disaster. It has escaped injury from all the many hurricanes that have visited the Southwest, and what were disasters elsewhere—at Sabine Pass, Last Island and Cheniere Caminada—were but ordinary storms here."

CHAPTER XIII.

GALVESTON HAD WARNING—OTHER STORMS.

MANY OTHER ISLANDS IN THE GULF HAD SUFFERED HER FATE.

TWENTY THOUSAND LIVES LOST AND $100,000,000 OF PROPERTY DESTROYED IN THE SAME WAY—MANY CITIES AND VILLAGES WIPED OUT BY FLOOD IN THE LAST TWO CENTURIES.

The destruction of Galveston by the storm of September 8 and 9 supplies further evidence that the low-lying sand islands fringing the shores of Mississippi, Louisiana and Texas for a distance of nearly 1000 miles were never intended for human habitation. The evidence has been secured at a frightful sacrifice of life and property, for it is estimated that more than 20,000 lives have been lost and nearly $100,000,000 of property has been destroyed on these islands by hurricanes. Island after island has been the scene of destruction, its population either wiped out or the island itself washed away. Galveston was frequently threatened before and warned of the fate, but the people believed that their breakwater and substantial buildings would be able to defy the storm. The Galveston horror is the twelfth of its kind in the sand islands of the Gulf. It is the worst in loss of life and property, because a city had been built on Galveston Island, but the other disasters were equally bad, or at least proportionately bad, and in most cases the percentage of saved was even smaller than at Galveston.

The record of the loss of life goes back to a time before the white man landed on these shores. When Bienville and Iberville, two centuries ago, occupied this country they found upon the island where they landed opposite where Mobile now stands the bones of hundreds of the aborigines. The French thought that the unburied bodies indicated a massacre, and so called the island Massacre Island. It is Dauphine Island today. It was a massacre, not by human hands, but by some mighty storm, which overwhelmed the Indian population and wiped it out of existence. It left a memory behind which the natives never forgot, and the Gulf islands were to them a haunted place never to be visited, in spite of the fact that they teemed with game. When the French settled there they found the islands wholly uninhabited. The Indians had learned what the white man has learned since, how unsafe they are for human settlement. Nor were the French long in learning this lesson, for within two years of their settlement on the Gulf coast, in the very beginning of the eighteenth century, their fleet went down in a hurricane in the magnificent harbor of Isle des Vaisseaux, now translated into Ship Island. Ship Island is almost identical in most particulars with Galveston Island, and it has been the aim of the people of Mississippi to build a great port there like the Texas city which would handle the commerce and free them from the control of New Orleans. The project has been popular for half a century, and a great deal of money has been expended on it. There is a splendid harbor on Ship Island, and by connecting it with the mainland as Galveson was connected in Texas a deep-water port could be obtained. A railroad has been built from the mainland into the interior of Mississippi, and all that is now needed is to connect the island with the shore by means of a pier or bridge. This has been repeatedly proposed, but never done, mainly for financial reasons. The destruction of Galveston will have a tendency to delay if not to kill the project, particularly as the Galveston storm washed away a portion of Ship Island.

Stretching from Mobile to the Rio Grande is a long fringe of islands differing very little from one another. Indeed, any of them might pass for Galveston Island, so alike are they, the only difference being that some of the eastern ones have trees, whereas west of the Atchafalaya they are generally bare and open, and therefore more dangerous, being exposed to the waves and wind. They range from ten to twenty miles long and from one to two miles wide. They are composed almost exclusively of sand underlaid by clay or quicksands, covered in part with a coarse, scraggy sea grass. They begin at Dauphine Island and stretch westward as follows: Petit Bois, Horn, Dear, Ship, Cat, the Chandeleurs, Breton, Bird, Grand, Timbalier, Caillon, Last Island (originally one island, but cut into two by the great storm of 1857), Marsh, Galveston, Matagorda, St. Joseph,

Mustang and Sadie. Cheniere, Caminada and Indianola are in reality islands, although technically not known as such, being separated from the mainland by swamps always under water.

At one time or another every one of these islands has been struck by a hurricane and depopulated and changed or modified in shape. They will change their appearance entirely in a night. As a consequence most of them are without stable population today, as in the Indian days, notwithstanding the game on them. A few charcoal burners and cattlemen live on Cat Island. The others are mainly frequented by fishermen, for they are fine fishing stations. Of the entire lot, perhaps, Chandeleur Island, as it is called, is the worst. It may have been an island once; it is an archipelago today. Facing southeast and acting as a breakwater to the country around New Orleans, it catches all the storms, and has been the scene of more wrecks than perhaps any other place in the country. Originally forty or fifty miles long, the storms and the ocean have cut it up into a score of islands, and its form changes with every hurricane. It has completely changed its character within historical times, and is apparently being washed away and likely to become in time a mere reef. The water now pours over it with every storm, and the island disappears completely from view, buried under the Gulf. Originally covered with wax myrtles, from which the creoles made their candles, hence its name (Chandeleur), it now boasts of nothing but marsh grass and a single palm tree which by some strange freak of chance has survived all the hurricanes.

Some years ago the United States government, on the persistent demands of the people of Louisiana and Mississippi, established its Gulf quarantine station on Chandeleur Island in what was supposed to be a very safe spot. In the storm of 1886, which may be called the Sabine Pass storm, since it was the Sabine Pass country that was overwhelmed that year, the quarantine officers had just time to get away from the station. When the physicians went back to look for their station they could not find a trace of it. The very site had disappeared, and a few battered pieces of wood picked up on the coast and supposed to be the wreckage of the quarantine station were the only remnants ever found. The station was never rebuilt, but moved to Ship Island. Chandeleur Island is wholly uninhabited today, and its sole occupants, except birds, are a species of wild boars, which seem to have some way of defying the elements. The other islands, Errol, Bird, etc., stretching to the mouth of the Mississippi, are but pieces of Chandeleur that have been separated from the main island by violent storms, which have torn it to pieces.

West of the Mississippi come the islands which have suffered most in the Gulf hurricanes, for many attempts have been made to settle them. Grande Terre, the first of these, was of old the haunt of that famous Louisiana pirate, Lafitte, and it is a curious coincidence that Lafitte, after being driven from the island by the United States federal authori-

ELEVATOR A.

ties, should have sought refuge in Galveston Island, where he flourished for several years, the island then belonging to Spain and being wholly uninhabited and without the pale of the law.

West of Grande Terre is Grand Isle, and immediately adjacent thereto Cheneire Caminada. Grande Isle has been visited by a dozen storms and severely ravaged by them, but while nearly all the property on the island was wrecked in 1893 and many lives lost, it escaped wholesale destruction, thanks to a grove of oaks planted many years ago, whose roots act as a sort of levee or protection to the land. Not so fortunate is Cheniere Caminada, lying just across the channel and only two miles distant. It was the worst victim of the hurricane of October 2, 1893. At that point alone in a fishing village known as Caminadville no less than 1150 lives were lost, and 1678 were lost in all, every one of the neighboring islands having suffered. The bodies of only a few of the dead were recovered. The great majority were swept out to sea, and many were found by vessels fifty miles distant from the shore.

West of these islands come the group of Tunbalier, Caillou and Last or Derniere Island. They were the victims of the Last Island storm of 1857, so named from the island which was the worst sufferer, just as the storm of 1875 was the Indianola storm, that of 1886 the Sabine Pass storm, that of 1893 the Cheniere Caminada storm, and that of 1900 the Galveston storm.

The Last Island storm was memorable because of the

large number of prominent persons drowned. Last Island was a pleasure resort at the time, and the hotel there was crowded with prominent Louisianians. The storm that destroyed it was like all the others in its origin and action. A violent wind drove up the water in the bays back of the island, piling it ten or twelve feet high there. Then it veered from south to north, driving the waters back on the Gulf with a force that swept everything before it and out to sea. The wind and the waves cut the island in half, and where the fashionable Last Island Hotel once stood is now a part of the Gulf of Mexico. There were only 284 victims of the Last Island storm, but they included the lieutenant-governor of Louisiana, the speaker of the State house of representatives and many others prominent in the political and social history of the State.

West of Last Island the islands are too low and soft for human habitation, and in consequence they have never been settled. They have gone under with every storm, but it has fortunately been without loss of life. Near the Texas line some twenty years ago a large number of farmers settled on Johnson's Bayou. It is what would be called high land on the Gulf, rising six feet above the water. The settlers planted orange trees, and soon had some of the best groves in the State, but in 1886 a tornado struck them and the settlement was annihilated and some 250 persons killed in identically the same way as at Last Island. The wind drove the water into the swamp back of the bayou, then changed from south to north and swept the land away into the Gulf.

A few miles from Johnson's Bayou is Sabine Pass, which met with a similar disaster. The water piled up in Sabine Lake and dashed down on the town, which is situated on a peninsula or island at the mouth of the lake. It was completely swept away, with great loss of life. There is a Sabine Pass today, the terminus of a railroad, and a great deal of anxiety prevailed in regard to it during the Galveston storm of the other day, but it is an entirely new town and some distance from the town wrecked in 1886.

Next comes Galveston, and beyond that is Indianola, which, although theoretically on the mainland, is practically an island. The hurricane which swept over the Gulf coast in October, 1875, struck Indianola just as the other storms have struck Galveston, Last Island, Sabine Pass, Johnson's Bayou, Cheniere Caminada and other exposed points. Indianola at that time was a town of about 4000 people and the terminus of an important railroad system. It was a rival of Galveston, and deemed a dangerous rival for the trade of Western Texas. It was situated at the head of a long bay on land like that at Galveston, only a few feet above tidewater, and behind it there was such a network of bays, bayous and small lakes that it was really an island. The water was backed up by a continuous wind to an extraordinary height and for many miles back of the town. Then the north wind drove it seaward with a force that was irresistible, and only three houses were left standing in the town. It virtually killed Indianola, which is smaller than it was forty years ago, and its harbor is completely ruined.

Such has been the experience that Galveston had before it. But in no wise daunted it has gone to work to build up a great city on the low sand spit that juts out into Galveston bay. It was not a bit safer than any of the other Gulf islands; indeed, it is not so safe as many of them. Cat Island is covered with trees, whose roots bind together the ground; Ship Island rises in places forty feet high. It is true that these are mere sand dunes, liable to be swept away in a storm, but they afforded at least a refuge from the water when the wind drove it over the island, and the islands are broader. Galveston Island is only five feet high at best above the waters of the Gulf, only one or at most two miles wide, practically useless, with a foundation of salt, clay and quicksand. But at this sand spit some 50,000 persons settled and invested $30,000,000 in building up one of the prettiest and most prosperous towns in the South. Galveston has for years boasted of being the wealthiest city per capita in the Southwest, and was proud of the fact that Strand street alone possessed twenty-eight millionaires.

But all this prosperity was built on a quicksand. The people of Galveston knew this, as did everybody in the Southwest, but as year after year passed and Galveston escaped ruin in the storms which desolated or destroyed neighboring islands a spirit of confidence was aroused that it would altogether escape; that the town was too substantially built, too well protected by breakwaters to be ruined, as the less solidly built Indianola had been. The grade of the principal streets was raised a few feet, and the pavements were deemed a further protection against the waves and likely to prevent the washing away of the ground. But there were some who doubted, who built their houses like the dwellers upon Lake Maracaibo, ten feet from the ground, mounted high on poles, so that the sea could sweep under them if it rose too high and not flood the floors. And at every Gulf hurricane there were anxious inquiries whether Galveston had got through it without injury.

There was good cause for these inquiries, for while Galveston escaped serious damage from these tornadoes, it was only because the storm struck somewhere else and Galveston did not get its full brunt, and in all the cases it had a very narrow escape. In 1857 the entire island was flooded, and the waters of Galveston bay and Gulf met over it so that it completely disappeared from view; but the town was then a small one, and the loss of life was inconsiderable. In the storm of October 3, 1867, Galveston again went under water, the Gulf pouring over it so that Mechanic street, the principal business thoroughfare, was six feet deep, and it then was on the edge of the storm and did not catch its full force. Again, in 1871, it was twice beneath the waters, first in June and again in September, one flood coming from the waters of the Gulf, the other when the water was piled up in the bay until it swept through the principal streets back to the Gulf of Mexico. In Ocober, 1873, and in September, 1875, and December, 1877, the town was again flooded.

Thus five times in ten years Galveston was swept by the waves and became a second Venice, all of its streets being from two to five feet under water. All of those storms were severe and did great damage, although Galveston caught only their fringe. But the storm of 1875 was by far the worst, and Galveston then escaped by only an hour, perhaps, the disaster which has visited it today. Had not the wind changed at the very moment it threatened to destroy the Island City the latter would have probably been swept into the Gulf with great loss of life. The storm did Galveston an immense amount of damage, and there were lives lost all along the Texas coast, but the city escaped a great catastrophe. A strong south wind piled the water up in Galveston bay until in Buffalo Bayou near Houston it reached a height of thirty-seven feet. Forty persons were drowned in and around Galveston. Morgan's dredging fleet was sunk, the government works swept away and incalculable harm done. Then the wind veered around to the north and all this immense mass of water was thrown back on Galveston Island. In twenty-five minutes it had cut the island in half, making a channel 250 wide and 25 feet deep at the east end, near Fort Point, and just beyond the built-up portion of the island. The land washed away like so much sugar, and it was evident that the entire island would be swept into the Gulf, but just as the new-made channel reached the city the wind receded again, the water was driven eastward and passed out through Galveston Pass. It was the narrowest of escapes, for fifteen minutes more of that north wind would probably have carried a hundred houses out to sea and drowned every occupant. The channel cut by the storm of

1875 still remains as a warning of danger to every one on the island, unless it was destroyed in the storm the other day.

Two years afterward, in 1877, another storm destroyed the government works at Galveston harbor, but the town escaped any very great injury. The storm of 1886, which destroyed Sabine Pass and Johnston Bayou, was the last serious one to visit Galveston, and again that town was flooded.

These storms explain to a large extent the present Galveston disaster. It bred a feeling of desperate confidence among the people that no storm could injure Galveston. When, therefore, the hurricane struck it on Saturday, instead of seeking places of safety, they shut themselves in their houses and waited for the storm to blow. They knew, of course, that the streets would be under water, but the streets had been under water so often before that this did not carry the same significance to them as it would to the people of other cities. But this time the storm, which had dodged around Galveston so often before, struck the island fairly and squarely. This confidence caused the great loss of life. At Sabine Pass and other places which suffered from the recent hurricane the people sought refuge on the higher ridges or congregated in the stronger buildings, but in Galveston they shut themselves up in their houses and were trapped like so many rats.

It will be some time before it is possible to determine what effect the storm has had on the island proper and on Galveston bay and the jetties. Apparently the island is less hurt than by the storm of September 17, 1875, when the southern portion of it was cut off, but it is not certain that it has escaped permanent injury, for it is covered everywhere by sea ooze. As for the bay and the jetties, upon which the United States government expended $8,000,000, to which Galveston owes so much of its present prosperity, only a careful examination can disclose whether they have been injured or not. Judging by the experience of Indianola, Sabine Pass and other places, the chances are that the whole character of the bay and the surrounding country has been changed by the storm. But it has proved once again that in their present condition the sand islands of the Gulf were never made for settlement.

BURYING THE DEAD ON SPOTS WHERE FOUND.

FROM PIER 33, LOOKING EAST.

CHAPTER XIV.

HOMES ON THE SAND.

WHAT THE STORM HAS TAUGHT THE PEOPLE LIVING ON THE GULF COAST.

Many hints have been offered with reference to the rebuilding of the homes in Galveston since the flood, but perhaps none is more suggestive than that sent to the News a few days after the disaster from Grand Isle, La.:

"It is the duty of your great State to see that her one great port be maintained to her manifest advantage. But here comes the main question, that of adequate protection both from the surf and from the back waters of the bay, from the crushing force of the breakers and the gradual or rapid rise of the overflow, which, in connection with the strong wind prevailing, float off their foundations the lighter and more exposed buildings. If they do not go to pieces in falling, they are hurled against the next with a crash, and both go down in a common ruin and are carried against the next and banked up against it, and the force of the current soon washes the sand from under the foundations of the latter and it goes down. And so the work of devastation goes on, until nothing is left to destroy, or the wind and waters abate and recede and go down. Such was the case on the ill-fated

Cheniere Caminada in 1893, and to a lesser extent here on Grand Isle. On the Cheniere, of 239 dwellings, only five remained on their foundations. Four were built on the back ridge, altogether inland as the land lays, and the fifth, on the same ridge, half a mile to eastward of the others and far more exposed, a small house, 20x25 feet, and light, and set on brick pillars of three feet in height only.

"But from all the circumstances one is inclined to look upon its preservation, with the sixty-three souls it sheltered, ranging in age from the suckling to the great-grandmother of eighty-eight years and bedridden, to a special Providence. The nearest wreck, more than 200 yards away, need not be considered in what follows. The others afford an illustration in method of building. They were all studding frames of 4x4, sills 6x6 inches, plates 4x4, sleepers, joists and rafters 2x6 inches, and with perhaps a dozen exceptions every house there was of the same material and its dimensions. One of these four, 39x41 feet, with an annex 20x39 feet, stood on brick pillars, of which there were thirty-four. three and one-half feet high, lifting the floor four and one-half feet above the ground. All had a double open fireplace chimney in the middle of the house—a good backbone when built of good material.

"This house had two feet of water over the floor, but it did not move, because it had, besides the chimney, two heart of live oak braces set deep into the ground at each corner, shouldered onto the sills at the top of the pillars and spiked into the sills. Had the pillars been knocked away by drift or undermined the braces would have held it up.

"The other three were on ground a foot lower, set on live oak blocks or posts, set into the ground from one to two feet deep and braced in a like manner, the braces being sunk from two to three feet. One had the floor five feet above ground, and the others two and one-half and three feet respectively. One (the second and lowest) resisted the shock and weight of another house of almost its own dimensions being driven against it during the height of the storm, while there were over four feet of water on the floor. But it withstood the shock and was but slightly damaged from coming in contact. True, these houses were not exposed to the full fury of the high waves, but they were subjected to a long swell coming over a low marsh where the back water was anywhere from seven to ten feet deep, and driven over the ridge in depths varying slightly from five to six feet deep, and being driven by a wind which tore houses of lighter construction to fragments over the heads of their inmates.

"Other houses in the same section standing on higher ground, and higher from the ground by from one to two feet, went down before the wind and waves like houses of blocks. Those which withstood the wind and sea from the east and southeast until the deadly calm set in at 10.20 P. M. went down before the death-dealing northwester between 11.30 P. M. and 3 A. M. of Sunday and were swept into the sea.

"Here is, I think, a fair illustration of the superiority of wooden pillars, or piling, well braced and properly protected at the surface of the ground to prevent the ravages of wood-destroying ants and the action of the sun and dampness by the application of not less than two coats of pine tar, well applied, boiling hot, with a coat of coal tar, likewise applied hot, over the other, thus forming a gloss which makes the wood impervious to the action of the weather for years. On sandy soils, such as compose the islands of the coast, brick pillars are a snare and a delusion, unless they are started from at least ten feet below the surface of the ground, built with good cement and of a thickness which will enable them to withstand the impact of the largest sawlog or whatever drift may be hurled against them by a tempest. From what I can learn, all, or nearly all, of that resident portion of your city so exposed to the sea was built upon brick pillars, on a bottom of sand, which even the rains are liable to wash out. Hence followed the great destruction of property and consequent loss of life. Still I do not lose sight of the facts—the overpowering force of wind, wave and current.

"Here on Grand Isle we had, in 1893, ample illustration of the superiority of deep-set piling foundations over brick pillars. The Ocean Club Hotel, an immense two-story structure, 360-odd feet long, almost an E in shape, was set up on five rows of 12x12 cypress piling, driven into the ground from eight to ten feet. Had the roof and upper floor been securely fastened down the damage to the upper structure had not been one-fourth as great. Note that the rafters and upper-floor joists were fastened down with eight, ten and twelve-penny nails, when nothing less than twenty-penny should have been employed. Owing to this fact not one vestige of the roof over the main building (dining-room and kitchen excepted) remained. With the exception of the Ls and part of the kitchen, the walls remained. They were of $1\frac{1}{2}$x12-inch pine plank, and reached from sills to roof, weather-boarded. But the foundation: Out of 180 under the main body of the building only four, on the sea front, were undermined, except those under each of the wings on that front, which were all dug out, and three were partially and one totally lifted out on the north side. The hotel was exposed to the full force of the current from the sea and the receding tide during at least eight hours, as well as the full force and wash from the sea from 8 P. M. until the wind changed to northwest, and then to the sea behind it. Had those who could afford it had such foundations under their costly homes, with the floor sufficiently raised, many lives would have been saved.

"Another illustration: The immense hotel then being constructed by P. F. Herwig, and set on brick pillars four feet high, went down. Had it had a foundation of piling, such as the Ocean Club Hotel, it would have stood. The latter is being rebuilt on the same piling, and can withstand another '93, should such another misfortune occur. We have learned the value of piling and braces and high floors. Will Galveston heed the lesson? Advice is cheap, and, however well meant, is little heeded. Overconfidence is a costly experiment, condemned by wind and wave.

"Braving the sea, as Galveston seems to have done by building almost to the edge of the breakers, without even the protection of a revetment, was a most lamentable want of foresight, without a bush or tree to cling to in case of disaster. Parsimony is downright criminality. Plant quick-growing trees of all kinds that will grow; put a girth of forest between the beach and your homes. Galveston has a great future before her, but she must protect herself against the recurrence of the tragedy of today. In a few more decades she will have a population of 5,000,000 or more Texans behind her, mainly dependent on her enterprising merchants and people. We would be proud to see Galveston again where she was two weeks ago in point of population and material prosperity—aye, and far beyond that. We may be too old to see it, but it will come if Galveston is brave and wise and wills it. Let her enterprising public men and women gird themselves for the work and prove to the world that they are superior to adversity, and not only will the world applaud, but success will crown their efforts. Galveston Island has existed a thousand years. Why not another thousand?"

PART II.

STORIES TOLD BY SURVIVORS.

16TH AND M.

CHAPTER I.

AS DR. CLINE SAW IT.

THE EXPERIENCES AND OBSERVATIONS OF THE GALVESTON WEATHER BUREAU CHIEF.

WARNED THE PEOPLE EARLY.

A GRAPHIC DESCRIPTION OF THE RISING OF THE WATERS AND THE DESTRUCTION OF THE HOUSE AND ITS INMATES.

A "storm wave," Dr. J. M. Cline, chief of the Weather Bureau at Galveston, calls it. He draws the distinction between a "storm wave" and a "tidal wave." The wall of water which rolled in from the Gulf upon the beach side of the city was a storm wave created by the hurricane, and not a tidal wave caused by earthquake influence.

As he sat at his desk and talked of the various aspects of the visitation Dr. Cline looked like a fit subject for the hospital. Two or three linen bandages about his head covered from view scalp wounds. One foot was poulticed and swathed, and when the doctor went across the room to get a chart he hobbled painfully. Now and then he referred to his assistant and brother, Joseph Cline, and another bandaged head was turned to give the answers.

Dr. Cline has been here eleven years. He knows Galveston's conditions from the meteorological standpoint better than any book tells them. Upon the roof of the five-story office building where the government weather service is located, there are the broken poles and twisted wires which tell of the ruin of observing instruments. But up to the moment when the storm swept away the apparatus the record was kept with fidelity to official trust.

It is a matter of congratulation that, great as was the loss of life, the awful total would have been much greater but for the duty to the people which Dr. Cline and his staff performed. There are walking the streets of Galveston today and engaging in the work of restoration 1000 people who owe their existence to the warnings which were sent during the early hours of the storm. Perhaps it should be thousands.

"All day Saturday," Dr. Cline said, "one of our force was kept continuously at the telephone answering inquiries. The water was rising and people were asking from all parts of the city advice as to what they might expect. Our answer to all was that things would be much worse than they were, and that it would be well to take all possible precautions. We warned those not in strong buildings to seek places of greater security. When I went out my instructions were to tell everybody that the worst was yet to come. Some who had not telephone connections came to the office and were told that the indications were alarming. At 2.30 in the afternoon I started to make observations of the swells. I met people streaming from the residence sections into the business center, carrying bundles and seeking safety. Of course, I did not foresee what was going to happen, but the observations pointed to increase of the storm, and we did all we could to put people on their guard. I am satisfied that numbers were saved by the notice they received and the prompt action they took to get down town."

* * *

With his records before him, Dr. Cline told of the earlier stages of the storm. The barometer marked 28.53 inches. The barograph indicated a barometer of 28.44 inches. These readings were before the storm reached its height. To the practiced eyes of the observers they foreshadowed extraordinary atmospheric disturbance. Dr. Cline says the readings are the lowest for Galveston of which he has recollection. Just before the great storm here in 1875 the barometer gave 29.25 or thereabouts.

"The tide commenced coming into the lower part of the city in the early morning of Saturday," Mr. Joseph Cline said. "It continued rising slowly all day. Between 3 and 8 in the afternoon the advance of the water was very rapid until the entire front of the city was covered from six to fifteen feet deep. The wind and the water carried away all buildings in the east, south and west portions of the city for several blocks inland. I believe that the storm was much greater a little west of Galveston than it was immediately in the heart of the city. The path was not wide. It extended but a few miles to the east of us. The city was a little east of the center of it."

Dr. Cline, the chief of the bureau, was in a position to note the storm wave and its effect. He lived on the beach side of the city, where now there is only a waste of sand, but where, before the storm, houses stood in large numbers. Dr. Cline started from the bureau about the middle of Saturday afternoon to make an investigation of the swells. But before he reached the beach he found himself getting into deep water, and with difficulty reached his house, three blocks from the beach. At this point the narrative leaves off its scientific character and becomes personal and tragic. In a few minutes the question with Dr. Cline became one of life-saving rather than of storm-observing.

"I had been here some years," he said, "and had built my house with special reference to possible hurricanes from the Gulf. I had made it, as I thought, strong enough to resist any storm that might come. The timbers were heavy. The attic was braced. The fact that the roof stayed until all went shows how well constructed it was. Forty people found refuge in my house as the storm rose. We felt confident that we would get through. My place did not go until all others in the neighborhood were down. It stood, in fact, until the ruins of the other houses, carried by the waves from the Gulf, struck it like battering rams. About 6.30 the storm wave, which has been called by many a tidal wave, came in. The great mass of water was four feet high, as nearly as I could judge. I was at my front door when this wave struck it and dashed it in. After this high wave came other waves, but not so heavy. The water continued rising, however.

When the wave struck my door several men were behind, bracing it. The door was torn from the hinges and the men were thrown backward. The water was knee-deep in my parlor. Then it went to the armpits. The rise continued after the wave came until the depth at my house reached fourteen feet. At 7 o'clock, after the storm wave, there was a lull of a few minutes. We thought the storm was going to subside, but soon it came stronger than ever."

In that renewal of the fury of the elements the ruins of other houses were driven against the Cline residence, and it went down in the waters. Of the forty people within, only eighteen survived. One of the lost was Mrs. Cline.

"As the house went to pieces," said Dr. Cline, "I had one of the children and my wife. My brother had the other two children. The timbers crushed in around us, and the roof settled over us. We went down into the water. I tried to release myself, but found I was held there. The water came over my face. I made up my mind that was the end, and opened my lips, or thought I did, with the idea that the water would enter my lungs and death would come quicker. All consciousness passed from me. The next thing I knew, probably only a few moments later, my head was through an opening in the roof and above water. I was holding the baby by one ankle. A plank was floating toward the little one, and about to strike. This seemed to arouse me. I put out my hand to ward off the plank. Then I struggled out and got upon some floating wreckage. My brother had broken through a window as the house collapsed, and had pulled the children through, all of them receiving cuts. My wife I did not see after the house went to pieces."

Then followed a period of three hours of instinctive struggle to prolong life almost without hope.

"We drifted about," said Dr. Cline, "shifting from one part of the floating wreckage to another. For two hours we did not see a house. I suppose we went out into the sea and came back. My three children and one woman who had sought refuge in the house were with my brother and I. A little girl eight years old came climbing over the timbers to us. We saw no other persons. After we had been afloat some time the current carried us within sight of houses. In the dim moonlight we could see them being knocked to pieces as the drift struck them. About 10.30 our wreckage carried us to Twenty-eighth and P streets, where we were able to get out and to reach land. During that whole night I seemed to feel no fear. I suppose I was beyond it. I felt cool and deliberate. Now, however, it is hard for me to command my faculties. It is difficult for me to think."

A remarkable thing Dr. Cline states about the eighteen out of forty who were in his house; that is, the large proportion of children who escaped. In a house one block west of Dr. Cline's a hundred persons took refuge. Only two survived.

"This was a hurricane," Dr. Cline said. "Readings all around the West Indies indicated it. For three days a storm of this kind had been predicted for the Gulf region. On Friday we had warning signals up. I recollect that one citizen came to the bureau early Saturday and said he had made up his mind to put his family on one of the steamships in the harbor, he felt so sure that an unusual storm was coming. I couldn't say that the season preceding was extraordinary. In July we had heavy rains, but not in August. Galveston frequently has hard rains."

Dr. Cline, in answer to a question, told some interesting facts about the wind.

"In the early morning of Saturday," he said, "the wind was from the northwest. It shifted from northwest to northeast, and became steady from the northeast about 2 P. M. The velocity ranged from eighteen to thirty-six miles up to that time. At 2.30 P. M. the increase was to forty-four miles. The increase continued until at 5 P. M. the register showed the velocity to be seventy-four miles. Fifteen minutes later the record gave eighty-four miles, and when the anemometer blew away at 5.17 it had just shown an extreme velocity of 100 miles. At that time the blow was steady from the northeast. It is estimated that the highest wind came later from the east and southeast, and that the maximum velocity was probably from 110 to 120 miles. The wind blew steadily at 9 P. M., but there is no record. None could be taken. It abated about 10 P. M., and shifted to south and southwest. The tide fell much more rapidly than it had come in. Comparatively no damage was done after 11 P. M."

This shifting of the winds accounts for people being blown out into the sea on floating ruins of houses, and then brought back to the island.

CHAPTER II.

A NEWSPAPER MAN'S STORY.

AN EDITOR DESCRIBES HIS IMPRESSIONS DURING AND AFTER THE TEMPEST.

If the story of the late storm is ever fully written out I think it will be found that it originated in the West Indies, its first place of contact with land being in the Island of Jamaica, where the damage, judging from the meagre accounts received here up to the present writing, was nearly as great as on Galveston Island. That it left death and destruction in its wake is undoubted, but prostrated telegraph wires prevented details from there. The next heard from the storm was on the coast of Florida, doing some damage at Tampa, and later damage reports come from Miami. Early Saturday morning a special to the News from New Orleans reported a severe storm on the Mississippi and Louisiana coasts, but said the wires were down and details not obtainable. An Associated Press dispatch later gave practically the same information.

The writer left the News editorial room at 3.45 Saturday morning. When nearing his home the water was knee-deep on Avenue M and Eleventh street. This was between 4 and

"KENDALL CASTLE" BLOWN THROUGH BAY TO TEXAS CITY.

THE HOUSE IN THE FOREGROUND WAS LIFTED OFF ITS BLOCKS
AND CARRIED FIFTEEN SQUARES.

5 o'clock. High waves were washing over the "Susie" track at Eleventh and N occasionally, but up to daylight there seemed no danger from the waters of the Gulf, and in fact up to 11 o'clock people believed there would be no severe overflow, as the fierce north wind was apparently attempting to drive back the waters of the Gulf. But the waves gradually increased in size, and the water was forced further up the island. About noon even the more confident prepared to get out of their homes, some of the more timid having gone as early as 7 o'clock. About 1 o'clock the writer's family left, his sons returning after conducting them to a place of safety.

Then came an experience which is hard to describe. Sitting apparently calmly by a south window, watching the angry waters of the Gulf as they came inward in solid waves with but little of the usual whitecap accompaniment to a storm; watching huge timbers, boards, chairs, tubs, barrels and other drift going Gulfward, only to be forced back with terrific force and strike the piling of the house with a thud which made the building tremble. Gradually the boards inclosing the basement of houses southwest were torn from their places, one by one, leaving three houses standing with bare piling. Then came a swaying of the houses and they went down into the Gulf, falling northward as gently as a mother would lay her infant in the cradle. But once "in the cradle of the deep," the angry waters seemed to take on fresh fury, and in almost less time than it takes to tell it the first house to go down was wrenched and torn into kindling wood. The other two held, perhaps, ten or fifteen minutes longer, and then were things of the past, even the slate roofs being torn to atoms.

Then two houses came floating Gulfward from M½ and were wrecked almost opposite the house. Meanwhile, the huge timbers of the Ninth street jetty were torn from their places and came floating westward. These were more to be feared than anything that preceded them. After they had apparently passed the house one huge timber floated back east, was lifted on a wave and hurled under the house, striking the brick chimney and crushing it like an eggshell. Then hope was gone. Shortly after the house went gently over to the northeastward and settled in the water. I intended to fight it out in the house, but when a foot of water came up, at the urgent request of my boys, I got out and slowly made my way through about two blocks of driftwood to where my family was located, on Fourteenth near Winnie. Then I attempted to go to the News office to report for duty, but after reaching Seventeenth and Avenue H gave it up. This must have been about 4 o'clock, and the storm was then steadily increasing. Slates were flying through the air from the roofs of houses, and bricks from chimneys flying in all directions. Wires from telegraph, telephone and street-car lines were flopping in the wind, with here and there a pole or a tree going down before the wind. The rain was coming down in torrents and cut like a knife when the east wind, which was then blowing, was faced.

Once back in the house, we moved across to another house on the corner of Fourteenth and Winnie. Here a few anxious hours were passed. The water gradually rose until the inmates were driven to the second floor. At times the house trembled like a ship in a storm. The culmination of the storm at that particular place, judging by the height of the water, was reached at 8.10 P. M. From that time on the water receded, and though there were strong gusts of wind, they came at longer intervals.

Some time after midnight someone came to the house and said the people were all ordered to the courthouse, as a telephone message had been received that a far worse storm than had been experienced was approaching. Most of the people left, some going to the courthouse, others to St. Mary's University, my family and self going to the courthouse.

Not until I started out to get something to eat had I the slightest idea of the wreck in the business district and north half of the city, and I do not think one out of twenty of the people not outside of houses realized the terrible devastation. I expected the wreck in the southern half, but believed, with others, that the damage would be slight in the northern portion.

Rumors of the death of people flew thick and fast during the early hours of Sunday morning, and one of the peculiarities of these rumors was that they proved untrue. The first two deaths reported to the writer as having occurred were a prominent physician and an equally prominent shipbuilder, both of whom are still alive and working faithfully to help those more unfortunate than themselves. On the other hand, some who were first positively reported as saved are now known to be among the dead.

Having gone through the storm of 1886 in the house that went down in the present storm, I had a fair chance to note the difference in the disturbances, as far as the Gulf beach was concerned.

In the storm of 1886 the wind came from the southeast, and tide and wind helped early to pile up the waters. Waves fifty to seventy-five feet high were plenty far out in the Gulf, and one looking at them expected to be completely overwhelmed when they reached the shore. But they broke in huge white caps to be formed into another wave of slightly less dimensions, breaking at intervals, until by the time they reached the house they had gradually decreased in size, though many of the white caps striking the house threw spray completely over it.

In the storm of 1900 the waves far out in the Gulf did not nearly reach the proportions of those fourteen years before. The white caps were comparatively light. The strong north wind caused a sort of misty spray to be driven southward from the tops of the waves, which, while not so high, had an appearance of solidity not seen in those of 1886. And while the waves decreased as they neared the shore, they seemed to gather angry energy as they approached the buildings, and where they struck them it was apparently with twice the force of higher but thinner ones of fourteen years ago. It seemed as though the opposition met by waters only condensed them to greater bulk, but did not stop their onward march.

Not being weatherwise or posted as to the theories of the weather bureau on the late storm, I will, nevertheless, advance the theory that there must have been a parting of the winds after the storm left Jamaica, and one current must have gone down the south Gulf coast and turned north again some distance south of Galveston Island. This wind caused the rise of the water in the Gulf. The other arm of the wind, with fully as great force, reached the east coast of Florida, crossed the State and ran along the coasts of Alabama, Mississippi, Louisiana and Texas to Galveston. The latter wind caused the great destruction of property, while the other was responsible for the loss of life. I can hear of few or no losses of life north of Avenue I, in the East End, except on the bay shore.

Some people have said, "How did you feel when your house went down in the storm?"

It is a question easier asked than answered. I was among the fortunate few who lost their houses early in the storm and before darkness set in. Up to fifteen minutes or less before the house went down I hoped that it might survive the storm. For three hours before it went I watched the waters patiently, mostly from the south windows, but, of course, had the restlessness natural to people who are waiting for a great crisis in the lives of themselves or those dear to them. To sit patiently still under the circumstances was impossible. A few moments' rest by a south window was followed by an uncontrollable desire to go to some other part of the house to see how matters were looking. Wandering from one point to another, a round of the house was made, and once more I found myself back at the south window to watch the main danger point. I do not think that I or any of my family could have been called excited. There was a restless, uneasy feeling among us all, but actually no fear.

When my wife left the house she fully expected to return to it when the storm was over. My boys were with her and my little girl, and for probably half an hour I was alone. During that time I was partly engaged in keeping the north and east doors closed. The wind blew them open several times, but did not break the hinges. When one was blown open torrents of rain poured in, and I remember thinking of the task the women would have in drying the floors and disposing of articles that had suffered from the water. From this it can be judged that even at that time I was not looking for a total wreck.

How did I feel? I was not excited. I was not in fear of my life. It seemed to me that what I regretted was the property loss and the struggle I would have to repair damages. But a total loss, a sweeping away of everything I had in the world, was not thought of. In fact, it is hard to realize now, a week after the storm. The mind cannot rest all the time on one's loss, and at times it seems when I want something at my house all I have to do is to go out and get it. My good wife last night caught herself the same way.

Speaking of the need of a shirt for Sunday, she asked:

"What do you want to buy a shirt for when you have three or four? Oh, I forgot, they were lost in the storm."

We have been housed safely, and it has seemed more like a visit than a total loss of property to her, except when she has felt the need of something that was carried away in the storm. As time passes and we begin to realize that all is gone, there is a desire to find something, even if it is of no value, when the wreckage is cleared away. My wife expressed a wish that the family Bible might be found, be it ever so dirty or torn. It contained records which could be nowhere else secured, and if a new one is purchased and the records again written it must be entirely from memory.

But though we lost all, we were among those families where no life was sacrificed in the storm, and in that respect were more fortunate than some of our neighbors and many of our friends.

The number of broken families in Galveston seems innumerable. As one walks the streets he meets friends of whom he had never thought, and the first greeting is, "Did you save all your family?" An affirmative answer brings out the remark: "You are lucky; many have lost not only their worldly goods, but their families." In many instances the reply is that your friend has saved his family, but has lost other relatives.

It seems that there is scarcely an individual in the city that has not lost some relative. Where the loss is not positive it is believed to have occurred because no news of the supposed dead ones has been received.

Previous conditions cut no figure in the display of heroism at Galveston. Walter C. Fisher, a druggist, started from his place of business for his home when the wind and water became serious. He was a man of not great physical strength. Mrs. Fisher had been a belle in her day, and her amateur performance of "Parthenia" is still remembered among the social events of the city's history. The Fishers had been married eight or ten years and had five children. They lived well out toward the beach. Mr. Fisher made his way through the water and in the face of the wind about half of the distance. As he passed a building where several men had stopped for refuge he was observed to be staggering from weakness. The men went out and brought him into the house. After a brief respite Fisher started again. As he left he said to the others, "If anything happens to me tell my wife I tried to reach her and died doing my duty." Walter C. Fisher did not reach his family. He was lost on the street. His wife and children perished when their home went to pieces.

A Mexican, whose name even is not remembered and who spoke but little English, conveyed his wife and children safely to the cotton mill. Then he went out to rescue others. He kept up this work until he had brought thirty-two women and children to the mill. During one of his absences his wife became frightened, got into the water outside of the mill and was drowned.

Galveston had a Grand Army post. One of the best-known members was Captain Seible, who had served in an

Ohio regiment. When the waters came up two attempts were made to get the old veteran to leave his home and to go to a more secure place.

"No," he said calmly, "it is better that I stay here and smoke my pipe."

His friends left the soldier sitting in his chair with his pipe in his mouth. The house went to pieces. When the storm was over comrades sought the ruins. The captain still sat in his chair, but the pipe was on the floor. The spark of life had gone out. A door had been thrown down with such force upon the veteran's head that a panel had burst out and the frame was about his shoulders.

In the early stages of the search for the dead a negro belonging to one of the parties came upon the body of his own child. The white foreman of the gang, sympathizing with the negro's grief, told him he might take the time to dig a grave for the little form.

"No," the negro said, "it is better for the living that all of the dead be burned."

He built the heap of wood, placed his dead child on it and resumed his work.

In the Galveston National Bank there was a lady bookkeeper. The water covered the street early in the day. At the closing hour it was several feet deep. The rise continued and the time soon came when the cashier and the teller were confronted with the question of getting that lady to a place of safety through water that was beyond wading depth. They solved it. A rope was fastened to the bookkeeper.

"You will have to excuse me, Miss," the cashier, who was a very polite man, said as he threw off his outer garments. Then they plunged in, the cashier swimming and dragging the bookkeeper to another building.

A spirit of calm resignation prevailed in Galveston during the worst of the storm. Occasionally there were hysterical screams from women and children or shouts for help from men. But the recollections of those who went through the horrors of the night go to show that most of the people who faced death were strangely quiet. In the upper story of one house forty-one people gathered and waited for the end. Scarcely a word was spoken for two hours. A little girl broke the silence by sobbing out:

"Mamma, how can I drown?"

At the Ursuline Convent, where 100 young ladies had assembled for the fall term of the school, the sisters led their charges in singing at times for soothing effect. Many colored people had sought refuge in the chapel, and as the storm increased they became demonstrative, exhorting and praying and singing with excessive fervor. The mother superior came into the chapel and, ringing the bell, induced quiet. Then she talked to the colored people. She warned them it was no time for excitement. She advised them to pray silently. She told them that if any of them wished to receive the sacrament in preparation for death it would be administered. A hush followed, and many of the people were baptized.

Four babes were born in the Ursuline Convent during the

storm. Their mothers had been brought in from nearby houses or floating ruins and had undergone travail in the cells of the nuns. One woman floated in a trunk to the door of the convent and was lifted out to pass through the sufferings of motherhood. As they came into the stormy world with scarcely one chance in a thousand to survive, it seemed at the time, the little waifs were christened by the mother superior.

It was with these sisters that Judson B. Palmer, general secretary of the Young Men's Christian Association, when his own home went to pieces, carrying down fourteen of the seventeen persons in it, found refuge. He floated against the side of the convent and was dragged in through a window.

CHAPTER III.

THE STORY BY FATHER KIRWIN.

THRILLING EXPERIENCES OF A CATHOLIC PRIEST IN THE GALVESTON STORM.

BELL TORN FROM ITS SOCKETS.

A FRANTIC HORSE KILLED BY A FALLING TIMBER—WHISKEY FOR THE MEN REMOVING BODIES—CARRIED TO SEA AND BACK AGAIN.

"Part of which I was and all of which I saw," might well apply to Father Kirwin's story of the storm. The tall, pleasant-faced priest of the Galveston Cathedral told in a calm, measured way of thrilling experiences through which he passed, and of the awful sights which came under his observation.

Of the half-dozen churches in Galveston the cathedral sustained the least damage. High up, visible from a considerable distance, the statue of the virgin still stands. But the bell was blown from its fastenings and tumbled down on the floor of the tower. The statue of the virgin was placed in its position soon after the great storm of 1875, and the act was prompted by that visitation. The massive bell hung in an open tower. It was not lifted out of open sockets, but was torn from strong fastenings. Looking from the windows of the parochial residence Father Kirwin saw evidences of the terrific force. The air was full of flying debris of every description. A frantic horse, he says, came dashing down the street. As the animal reached the front of the residence a heavy timber struck him and he went down. As the storm grew more furious the inmates of the house believed the end was near for them. Bishop Gallagher turned to Father Kirwin, and, indicating the several assistants, said: "Prepare these priests for death."

* * *

"We knew little of what had happened," Father Kirwin said, "except in our immediate vicinity. Very early in the morning my assistants and I started out to go through the parish to see what we could do. The cathedral parish extends across the city. We had not gone far before the general destruction began to impress us. When I got back to the house from a hasty round I told the Bishop I thought that at least 500 persons must have perished. I had not seen the beach side of the city or the western section. Going down to the wharf where several of our people were taking a boat to cross the bay for the purpose of giving to the world information, I said to one of them: 'Don't exaggerate; it is better that we underestimate the loss of life than that we put the figures too high and find it necessary to reduce them hereafter. If I was in your place I don't believe I would estimate the loss of life at more than 500.'

"You will see from this how little we realized on Sunday

morning of what had occurred. Now, after having seen every part of the city, and after having had to do with the collection of the bodies, I am convinced that over 5000 perished, and will not be surprised if the number reaches 7000."

"Only after I had made an extended examination of the city, continuing until late Sunday, did I begin to appreciate what had really happened to us," Father Kirwin went on. "As I was coming back down-town I met Mr. Morrissey. 'Father,' he said to me, 'we'll never be able to gather and bury all these people. There is only one thing to do, and that is to put them on barges, take them out to sea and sink them in the Gulf.' It was decided that this course should be pursued. Bodies were collected from the streets and from places where they were partially uncovered. Some of our best men took the lead in this, to set the example. They went right out and helped pick up the bodies. But hard as we worked, the more there seemed to be. It soon became so that men could not handle those bodies without stimulants. I am a strong temperance man. I pledge the children to total abstinence at communion, but I went to the men who were handling those bodies and I gave them whiskey. It had to be done. Monday night came. The barges were loaded. Out on the wharves and up the street were the floats still loaded. I heard one of the men in charge say, 'My God! don't bring any more.' Those who had been working all day were in no condition to continue. An armed guard brought up fifty negroes. The latter were driven on the barges, and the guard went with them. The barges were taken out into the Gulf and remained there all night until it was light enough for the negroes to fasten the weights and throw the bodies overboard. When the barges returned those negroes were ashen in color."

* * *

All of the Catholic institutions of the city suffered, but utter annihilation overtook the Catholic orphan asylum.

"I have been out to where the asylum stood," Father Kirwin said, "and have tried to find traces of it. There is absolutely nothing, unless it be a few scattered bricks. The asylum was not far from the beach. It was in that part of the city which was swept clean. The structure was large and strongly built. We have been able to find scarcely any part of it. At a distance of two miles down the island the other day I came upon the contribution box, which was in the parlor of the asylum. There was still upon it the inscription, 'Remember the Orphans.'

"Ten sisters were in the asylum. One of the community survived. She had gone down the island in a wagon and found refuge in a family. The others were lost. All of the children perished, with the exception of three little boys, who crawled through a transom, climbed upon some floating material and drifted to a place of safety. The orphans numbered about 100. Yes, the story is true that we found the bodies of one of the sisters with several of the children fastened tightly to her. She had evidently tied the children together and to herself, intending to save them or go down with them if the asylum went to pieces. Only three of the sisters lost have been found and buried."

"I had a very strange experience a day or two ago," Father Kirwin said, after a pause and with a deepening of tone. "A negro came to me. He drew out of his pocket the rosary and cross of one of the sisters. He began to cry. 'Father,' he said, 'I found her. I took this from her body and I buried her. I have got the grave marked, and I will take you to it.'

"One of the sisters, it is supposed, was carried across the island and then across the bay. We have the report from Virginia Point of the burial there of some one in the garb of a sister. The asylum was conducted by the Sisters of the Sacred Word."

Father Kirwin was asked to tell of two or three of the most marvelous escapes which had come to his knowledge and in which he placed credence.

"Well," said he, "there is the case of Ben Meyer, the butcher. Meyer was carried out to sea on Saturday. At 10 o'clock Monday morning he was met down the island walking back to town. Saturday night and Sunday he had drifted about in the Gulf."

"Ayers of the custom-house had a very strange experience," the Father continued. "When the house he was in became unsafe he took to a door. He was floating in the darkness when he came upon two children clinging to boards. He pulled the children to him and then discovered that the door would not carry them and him. He managed by swimming and pushing to get the door against the side of a stable.

BACK OF GRAND OPERA HOUSE.

FIRST BAPTIST CHURCH, CORNER TWENTY-SECOND AND AVENUE I, EAST SIDE.

IN THE BUSINESS DISTRICT. BIG RETAIL STORES ON MARKET STREET.

DEBRIS WORKED FROM GALVESTON INTO TEXAS CITY.

Above him was an opening into the hay mow. Ayers succeeded in boosting the two children into the mow. He clung to his door until the water went down, and toward morning made his way into the city, forgetting all about those children. The next day, recalling the incidents of the night, he remembered why he had pushed up beside the stable and why he remained there until the water went down. He returned to the stable and found the children. Not until then did he learn that they were his sister's little ones."

Father Kirwin tells some curious facts about the effects of the storm upon the churches. St. Patrick's had a tower 210 feet high, erected at a cost of many thousands of dollars, and only recently completed. This tower fell so that it lay across the middle of the church, cutting roof and side walls to the ground. The altar, however, was scarcely disturbed, and there every morning since the storm the priest of the parish has held service.

"The Ursuline Convent," Father Kirwin said, "was one of the most beautiful structures in the country. The interior adornment was very fine. A school was conducted there and quite a large number of the girls had arrived. The fall term opened a few days before the storm. Sixty sisters and the pupils were in the building, which stood in large grounds surrounded by a massive wall eight or ten feet high. This wall was leveled almost its entire length. Masses of the ruins of houses were washed into the yard, carrying people who had been in their homes when they collapsed. As cries for help were heard the doors of the convent were opened at the risk of those inside, and people were dragged in from the storm.

In more than one case persons rescued recognized the voices of relatives on the outside as they were borne on the floating ruins to the front of the convent. A woman said, as a shout was heard, 'That's Jim's voice,' and so it proved to be. When the man was pulled from his raft and into the convent he was recognized as the woman's husband. The Ursuline sisters moved the girls from one part of the building to another. At times they led the girls in singing to keep them composed. Along one side of the convent yard the ruins of houses and household contents are piled up thirty feet high. This great mass has not yet been overhauled for bodies."

* * *

One of these strange experiences was at the Sacred Heart Convent. The building sustained serious damage, but the community within was spared.

"A statue of the Sacred Heart," Father Kirwin said, stood in the chapel. It remained in place throughout the storm. In front of it the sisters and the children gathered. 'As long as the statue stands,' the mother said to them, 'we are safe.' The sisters tell me that those who were present remained motionless and silent, with their eyes riveted on the statue, while the storm raged. Although the walls were partially demolished, not one of those worshipers was hurt."

* * *

Father Kirwin saw some conditions which give vivid impressions of the destruction of human life.

"In the western part of the city," he said, "there is a place where a small railroad bridge crosses a bayou. When the waters went down forty-three bodies were left hanging upon the framework of that bridge. They were in the strangest positions and presented a spectacle which was horrifying.

"There is a place in the western portion called Heard's Lane. A citizen named Heard built a dike and set out on the raised ground salt cedars. In those trees lodged over 100 bodies. The horror of such spectacles was increased by the fact that all of the bodies were stripped of clothing. I know of but one body being found which was not naked. That was a fireman. I do not think that we have found nearly all of the dead. My belief is that as the great masses of ruins are cleared we shall discover many more. In our parishes we are now at work trying to make up an accurate list of the dead. My belief is that not fewer than 1000 members of Catholic families died."

"The people," Father Kirwin said, "do not realize the losses of relatives. They are still stunned. A curious instance came under my notice. Working for me near one of the institutions during the early search for bodies was a man whose manner did not indicate that he was suffering deep grief. But I saw that every time a body was found this man dropped his tools, went to the place, got down by the remains and examined the mouth. After he had satisfied himself he returned to his work, took up his tools and proceeded as if nothing had happened. I inquired about it. The man had lost his wife. He knew the structural form of her teeth, and he was trying to find her. But he did not show grief in any of the usual forms. You will hear people talk without emotion of the loss of those nearest to them. We are in that condition that we cannot feel."

CHAPTER IV.

STRANGE SCENES IN A CONVENT.

THE DOORS THROWN OPEN TO ALL COMERS—FOUR CHILDREN BORN—SISTERS OF MERCY.

The Ursuline Convent and Academy, in charge of the sisters of St. Angeli, proved a haven of refuge for nearly 1000 homeless and storm-driven unfortunates.

The stories of this one night within the convent walls read like the wildest dream of a novelist, but the half can never be told. Every man, woman and child who was brought to the convent or drifted there on the raging torrent could tell of an experience that would be well worth its publication. No one was refused admittance to the sheltering institution on this night of nights. Negroes and whites were taken in without question, and the asylum was thrown open to all who sought its protecting wings. Angels of mercy went through the army of sufferers, whispering words of cheer, offering what scant clothing could be found in this house of charity, and calmly admonishing the terror-stricken creatures to have faith in God and pray.

In contrast to this quiet, saintly and loving spirit of the nuns, the hundred or more negroes grew wild as the storm raged. They shouted and sang in true camp-meeting style, until the nerves of the other refugees were shattered and a panic seemed imminent. It was then that Mother Superior Joseph rang the chapel bell and caused a hush of the pandemonium. When quiet had been restored the Mother addressed the negroes and told them that it was no time or place for such scenes. If they wanted to pray, she said, they should do it from their hearts, and that the Creator of all things would hear their offerings above the roar of the hurricane, which raged with increased fury as she spoke.

The negroes listened attentively, and when the saintly woman told them that all those who wished to be baptized or resign themselves to God nearly every one of them asked that the sacrament be administered. The panic had been precipitated by the falling of the north wall of that section of the building in which the negroes had sought refuge. Order was restored through this woman's determination and presence of mind.

Forty sisters and forty-two boarding scholars from all parts of the State were in the convent when the storm broke. Soon the storm-driven people began to arrive in crowds of twenty and thirty. They were taken in through the windows, some were dragged through five feet of water into the basement, rescued by ropes from tree tops and snatched from roofs and other wreckage as it floated through the convent grounds.

Within this religious home four mothers who were rescued from death lay on cots in the nuns' cells and four little innocents came into the world. Mother Joseph, in speaking of this incident, said she believed it was the first time in the history of the world that a babe had been born in a nun's cell in a convent. As no one expected to live to see daylight, it was voted that these children should not leave the world they had just entered without baptism. Regardless of religious belief, the parents felt that the good sisters should administer the baptism, which in time of great danger is permitted without the presence of a clergyman.

Mrs. William Henry Heideman's home went down and was swept away. She clung to the roof and lost all traces of the other members of the family. The roof struck some obstructions and Mrs. Heideman was hurled into a floating trunk, where she suffered agony until the trunk drifted against the Ursuline Convent walls and was hauled into the building. Meanwhile, in a tree just outside, a young brother of Mrs. Heideman battled with the storm, clinging fast to a tree limb. While fighting for life he heard a cry. Reaching out one hand, he caught the dress of a child. He was Mrs. Heideman's little son. A few minutes afterward a wrecking party sent out from the convent in response to cries for help found the young man and his nephew and brought them to the institution. The reunion of a part of the family followed a few minutes later.

Another family reunion in the convent was the meeting of Mr. and Mrs. James Irwin, who were swept from their home. Each had mourned the other as lost. When Mr. Irwin was rescued his only garment was an old corn sack. The only dry garment to be found in the institution was a nun's garb. Mr. Irwin received it and wore it thankfully. During the long hours of that night of horrors he went about the building rendering valuable assistance to his fellow-men. Mr. Irwin left the institution still garbed as a Sister of Charity.

"I can't help laughing when I think of it," said one of the nurses in the Sealy Hospital in talking about the storm. "You know we have a large number of invalids in the hospital. Many of them are very old. They are what we call confirmed invalids. I do not know that I ever knew one of them to walk a step during my two years' stay at the hospital. They require constant care and attention. When the storm came these lifelong invalids leaped to their feet with the agility of 15-year-old boys and dashed off upstairs. One of them knocked me flat on the floor in his haste to get to the second landing. They walked around with sprightly gait as long as the storm lasted. When it was over they collapsed into a helpless condition. They are worse now than they were ever before the storm."

A. J. Selkirk saved his little ones in a peculiar way. When he saw the house was going he went to the roof and, nailing a cracker box on it, placed the babies in it. The roof floated away and the little ones were rescued. Not a member of either branch of the family was lost.

During the civil war a powder-house was erected in what is now the west end of the city. This house had brick walls three feet thick and was considered the strongest structure on the island. Just across the street were three flimsy frame cottages. When the hurricane and the rising tide drove the people from their houses the more thoughtful of them rushed to the powder-house, which years ago had been converted into a residence. The less thoughtful or more panic-stricken sought refuge in the flimsy cottages. Every one in the powder-house, some two score, perished, and all in the cottages survived.

Twelve authenticated cases have come to light when men who had achieved safety risked their lives and lost them to save others.

With his wife in his arms, a man living in the East End went down with the flood. The waves broke over him, and when he rose to the surface his loved one was gone. He struck out blindly and tried to swim. He floated for two hours. At length his feet touched land. He waded to the shore and began trudging up the beach. He had gone a mile, perhaps, when he heard a woman's voice calling for help. He rushed through the waves to her rescue, caught her in his arms and swam to shore. Then he worked over her with all his might for thirty minutes. Finally she recovered. He watched at her side till daylight. Then they walked to a house near by. He had saved his wife.

CHAPTER V.

SOME PERSONAL EXPERIENCES.

VIVID SKETCHES OF STRUGGLE WITH THE ELEMENTS ON THE NIGHT OF THE AWFUL STORM.

A volume could be written of the personal experiences of the Galveston flood. Thousands met death bravely fighting for life, but many thousands more were saved, some of them after passing through indescribable danger and many hairbreadth escapes.

Mr. F. T. Woodward of Dallas spent a thrilling and memorable night in the Grand Central Station. He furnished the following graphic description of his experience:

At about 8 P. M. the wind, which for several hours had been blowing a steady gale, increased in violence, and signboards and awnings were torn from their hangings and whirled in the air like chaff.

In company with about 150 others, I was in the Grand Central Depot, which, standing as it does, isolated and alone, was exposed to the full force of the hurricane, and the first strong gust was followed by a sound of shattering glass.

Several of the windows of the general office overhead had

given away under the almost irresistible pressure of the storm.

This was the beginning of a night of terror—of seven hours of mortal dread.

The storm continued to rage with unabated fury, and the roar of the wind was accompanied by the sound of crashing glass as one after another of the many windows was torn from its fastenings and shattered against the brick walls of the building or upon the sidewalk below.

Women clasped their children in their arms as though they expected to be torn asunder the next moment.

Men began to scan the pillars and partition walls supporting the floor above and to take up such positions as seemed to be most conducive to safety in the event the huge building was razed by the mighty breath of old Boreas.

The crashing of glass was soon followed by a sound of ripping and tearing, which was clearly and distinctly heard above the almost deafening roar of the storm.

Section after section of the tin roof which covered the awnings and a portion of the building was rolled up like sheets of parchment and hurled hundreds of feet away.

By this time the roar of the wind and the sound of falling glass and tin was almost deafening.

To add to the terror and confusion the electric lights were suddenly cut off and the great building was left in total darkness, except where the trainmen with their lanterns stood.

It was then that many moved toward the main entrance of the building with the evident intention of seeking other quarters, but they were checked at the door by the blinding sheet of water which was being driven by the wind with mighty force, and which lay between them and any place of refuge.

They appeared to hesitate between a choice of being drenched by water and possibly struck by a flying section of roof and remaining in the depot until the end.

But the question was soon settled.

Even as they looked, the roof of the Grand Central Hotel was torn off, many of its inmates rushing into the street.

Almost simultaneously a wail went up from the people in the Lawlor Hotel as the big skylight on top was torn loose and fell crashing down the shaft, bearing pandemonium in its wake.

This seemed to satisfy those in the depot that no haven of safety could be found, and that they might as well await results in dry clothing, and they returned to their chosen corners in the several waiting-rooms near the dimly-lighted lanterns placed here and there, apparently resigned to their fate and determined to make the best of the situation.

But just then in connection with the roar of the wind, the crashing of glass and the flapping and pounding and tearing of tin, a new sound was heard—a horrid, rumbling, deadly sound.

It was that of falling brick!

Everyone stood crouched, prepared to leap to either side as the occasion might require.

Everyone realized the gravity of the situation, but no one made a sound.

There was no shrieking, no fainting.

Many women were there, and every one of them stood the ordeal with such fortitude as to lend courage to even the faintest-hearted man.

Even the babies were mute and clung to their mothers' necks in breathless despair.

Near and nearer came that awful rumbling!

A shower of brick and mortar fell in the rear of the ladies' waiting-room.

Nothing remained of the tin-covered awning.

Few, if any, doubted that the end had come and that in another moment all would be buried beneath the ruins.

But suddenly the sound ceased.

The brick had fallen and the lower story of the building remained intact.

It was soon learned that the entire wall stood unbroken, and that the fall of brick and mortar was but the collapse of several large chimneys surmounting the top of the building.

As soon as this became known the effect upon the awestricken mass was electrical.

Men lit cigars, women chatted and laughed, and though more chimneys fell, more glass was shivered and the loosened tin on the roof continued to pound furiously until nearly 3 o'clock in the morning, there was no more panic, and all felt that the building would withstand the fury of the storm.

And it did.

But an inspection in the morning revealed the fact that it was badly shaken and greatly damaged.

As my train left Houston shortly after daylight—nine hours late—nothing had been learned as to the havoc of the storm in other parts of the city.

But a number of lives were lost, and the damage to property is almost incalculable.

Along the road north of Houston scenes of devastation and distress such as will not soon pass from me were witnessed.

Buildings had been torn down, and the material of which they were built scattered over the ground for miles and miles

Trees had been pulled up by the roots and denuded of their branches.

Fields that had been smiling the day before with all the great fertility of this record-breaking year were bare, the plants having been grasped by the ruthless hurricane and scattered far and wide.

Hundreds of head of cattle had been killed.

Adelbert Beecher, a Pullman car conductor running to Galveston over the Missouri, Kansas & Texas Railroad, reached Galveston on Saturday morning and was there throughout the storm. Mr. Beecher says:

"But when the character of the storm is considered it is wonderful that anything in its path escaped death or wreckage. I was in the St. Louis cyclone, but it had but a mite of the destructive power that raged over Galveston for three hours. The wind was only a part of the wrecking force of the storm. The measureless waters of the sea were its ally and undermined and churned into bits what it failed to overturn.

"When I reached Galveston Saturday morning the indications of the hurricane were in the air, but there was no warning that it was to have sufficient strength to wreck the city. After I had registered at the Tremont House and changed my clothing I went down to the beach. The breakers were bounding high, and even then—it was about 11 o'clock—the waters were beginning their march on the city. At 3 o'clock in the afternoon I again walked down to the beach. The water had made considerable progress on its march, for the gauge showed that it was three feet higher up the beach than it should be.

"An hour later the threatening clouds began to pour out a deluge. It was such a rain as one sees but once in a lifetime. It fell in sheets and the vision could not penetrate it more than twenty yards. At the same time the wind increased its velocity from forty-five miles to seventy-five miles an hour.

"At 5 o'clock the situation began to take on a serious aspect in the minds of the people. The water had made its way through all the streets of the city, and the velocity of the wind was rising. It was perhaps blowing eighty miles an hour at that time. At 6 o'clock the water was even with the street curbs. It was as black as midnight out of doors.

"It was at this hour that the storm broke on the city in all its fury. The wind suddenly became cyclonic in its velocity, sweeping at the rate of 125 miles an hour. The rain fell in great spouts, and the roar of wind and water was deafening. Two hundred persons shut off from their homes sought refuge at the hotel. The guests of the Tremont House sat secure in the lobbies. They knew the storm was on, but little dreamed of the ruin it was working. They told jokes, laughed, danced and made merry, while death held a carnival in less fortunate sections.

"In a few moments the water found its way into the lobby. In an instant all appearance of revelry was banished and fear—the kind that men never forget—took possession of the mirth-makers. Once the water made an opening in the lobby it poured through in streams. In an hour it stood three feet deep where the guests had been sitting, and was still rising at the rate of half an inch to the minute. Above the roar of the storm could be heard the crash of falling buildings, and the hotel, built solidly as any structure on the island, shook from the assaults of wind and water.

"At 9 o'clock the velocity of the wind fell to seventy-five miles an hour and the water had not risen any for twenty-five minutes. It was then five feet deep in the lobby. Twenty minutes later the water had fallen three inches and the velocity of the wind still further decreased.

"When Sunday came the sky was still cloudy and the wind was blowing in from the sea with considerable velocity, but it was clear that the storm had passed. The scene was one of desolation and despair.

"Not a house in the city had been untouched. Few were left even habitable. The whole city was a vast swamp, the waters having filled the streets with slime. There was nothing wholesome to eat. The only provisions left were canned goods, and they were spoiled. There was no water to drink. Famine and pestilence walked in the track of the storm.

"All provisions in the city were confiscated on Sunday by the authorities, and the survivors were put on rations. Dead bodies of men, women, children and animals of all kinds lay promiscuously about the streets and on the beach, but no effort was made to bury them. With the coming of night ghouls made their appearance and carried on a profitable business in robbing the dead."

"When did you first realize that you were in danger?" That, ordinarily, would seem to be a foolish question to put to a man who had escaped death as it rode on the storm, and yet it was not a foolish question, but the natural one. For the Galveston people had for years and years and years argued out the question of the danger attending the living on the island. True, Indianola, awful even now in memory, stood out as an alarm to those who live down by the sea. True, there had been storms and storms in Galveston. True, there were people on the great mainland who contended that wind and water would bring disaster to Galveston whenever the two acted in concert and from the right direction. But the answer to the Indianola alarm was that the situation of that unfortunate town exposed it to a storm fury; that it was a fair mark; that it was almost level with the water, and all that. The fact that there had been storms and storms at

A REFUGEE CAMP.

ONE OF GALVESTON'S PRINCIPAL SCHOOL BUILDINGS—MANY
PERSONS WERE KILLED IN THIS BUILDING.

BACK OF MALLORY WHARF.

NEAR ELEVATOR "A."

Galveston only confirmed the people in their security. For, as each had passed away without carrying any great number of lives with them, why should not this do the same?

As to the people on the mainland who had prophesied disaster, why they were merely timid and ignorant people. Therefore the question of "when did you realize that you were in danger" was a reasonable one. And the answer was the same in nearly every case. There might have been a difference as to the moment when these people, penned like rats in a cage, first felt the terror of impending death, but invariably the answer was that the storm was almost at its height before the realization came. In many cases only the falling houses brought the realization.

One little girl at a grocery store out on Avenue P, from which street to the Gulf the storm swept the island like a broom, answered me: "Mother and my eight little brothers and sisters were upstairs, and I went down to see what the water was doing in the store. You see, we live upstairs over the store. My papa is dead a long time ago. When I went down my brother went with me, and the water was half way up the counter. But that didn't scare us, because we have seen high water and heard the winds before. Well, we went back, and in a few minutes we were down again. Then the counter was floating. Brother said not to tell mother, but I did. Then we saw a house tumble down and we heard people crying. We got scared then and me and mamma prayed. We prayed that one of us would not be drowned if the little children were not drowned, because one of us would have to be their mother."

The maternal love was uppermost. But the love of that little girl for her little brothers and sisters, as she told me the story in her simple way, passeth in greatness all understanding.

"When did you think you were in real danger?" I asked of a merchant.

"Not until Ritter's saloon went down and I saw the waters rapidly climbing the walls. We had passed through the terrible storm of 1875 and had lived. Since then the island has been raised five feet or more. Why should we not have felt easy? But when the wind and waves began to show their fury, when I saw these extra five or more feet covered by a raging torrent which raced hither and thither, I felt that the end had come. Up the waters came about the fence, up they came and covered the hedge, up they came and knocked at the door. Yet I still thought the end would be reached. We had been told that the height of the storm would be at 9 o'clock. At 5 and 6 and 7 the waters continued to climb and the wind to take on new strength. At the last hour they were at the door. What must come, then, at 9? My heart fell then. I had peered out of the window and saw the dreadful enemy assault the house. Then agonized people were heard. It was dark and the spray sped in sheets. Yet it was light enough to see now and then. People in boats and wading came along. Their houses were gone. Mine rocked like a cradle, and I felt the end had come."

Thus said another man:

"What were your feelings?"

"Nothing but that of complete resignation. I have read much in books of the tableaux of the past appearing to the human mind on the eve of man's dissolution. In no instance have I found that the survivors of this terrible thing remembered the past. Some were frightened and simply shrieked and laid hold of anything that would relieve them from the embraces of the water. Some were frightened and prayed for mercy. Some were frightened into dumb resignation, partaking of dumb indifference."

There is one little hero who is different from the others. It was not life he tried to save. It was an inanimate object, but the story as told by Hunt McCaleb sent a thrill through every listener.

McCaleb was adjutant-general in General Scurry's administration of the city under martial law. He was in personal charge of the headquarters office, and was the busiest man in Galveston ten days after the storm.

During the Spanish war McCaleb was lieutenant-colonel of the First United States Immunes. In his regiment was an Italian boy, only fifteen years old at the time, who had managed to enlist in some way. His name was Adolf Biondi. Colonel McCaleb took an interest in the boy, and found him so earnest and faithful that he offered him a home after the regiment was mustered out. Young Biondi went to live in the colonel's beautiful home in Galveston near the south beach. Colonel and Mrs. McCaleb became much attached to Adolf. There were no children in the family.

"I saved my wife," said Colonel McCaleb in telling the story, "but we lost everything we owned. I have not a cent in the world. Our home was beautiful. We had such pretty trees and vines in the yard, and flowers, and—oh, the devil! There's nothing of it left. If I had lost my wife I believe I should have quit for good.

"After the storm I deemed it my duty to offer my services to the city, and I was busy from the first. I quite forgot little Biondi in the stress and press of terrible things. You are here, and you understand how that could be.

"About two days after the storm young Biondi, with about a third of a suit of clothes on and patches and bandages enough on his head and hands to make a shirt, limped into headquarters and stood at attention. He saluted gravely, just as he was wont to do during the war.

"I was surprised to see the boy alive, but very glad, for he is brave and good.

"'Well, what is it, Adolf?' I asked.

"'I held on to it as long as I could, Colonel,' said the boy, with tears coursing down his cheeks. 'I tried to save it, but it was no use, and I had to let it go, and it got lost, and here I am, and it's gone.'

"I thought the boy had lost his mind. 'What do you mean—what on earth are you driving at, Adolf?' I inquired.

" 'It was the flag, Colonel; I tried to save our flag, our regimental flag. I tried to save it, but it was torn from me.'

"You may know that I was deeply touched. Being urged to tell about it, the boy said that when the house went down he grabbed the flag that we had brought home from the war. To him it was the most sacred relic, and he thought that to lose it would be a disgrace.

"He floated out amid the wreckage clasping the folds of the flag to his breast. He had to swim, to dodge flying timbers and slate and brick, but for a long time he clutched the flag. It became torn, tangled in the whirling debris. It dragged him down under timbers, and he choked and strangled, but he held on to the flag. At last the hurricane snatched it from his hands and he was cast upon a drift, bruised and battered, but a true soldier boy still.

"And now he cries because he could not save the flag."

An official in the Treasury Department at Washington received two private letters from F. W. H. Whitaker, chief engineer of the revenue cutter Galveston, which braved the storm at Galveston.

The first letter of Engineer Whitaker was written September 9 before the storm had hardly abated and when its effects were not fully known. In this letter Mr. Whitaker tells of the rescue of thirty-four persons by Second Assistant Engineer Root and a few of the crew of the Galveston. The story of this rescue describes one of the most heroic deeds of the terrible catastrophe. When the storm was at its worst

Mr. Root and a few of the men went up one of the streets of Galveston in the whaleboat of the cutter. The water was then from five to eight feet deep in the streets. Thirteen men, women and children were taken from houses and placed in the boat, and the return to the Galveston was begun. The wind was so furious that the boat could not be moved an inch by means of oars. Mr. Root thereupon procured a rope and sent one of his men ahead with it. Sometimes the man had to swim, and again he would wade a short distance, with his mouth barely out of the water. He would fasten it to a telegraph pole or whatever held, and the boat would be pulled forward by the rope. When the boat was made fast with a small rope the larger rope was sent ahead again, the man who took it proving himself as great a hero as Engineer Root. In this tedious way the Galveston was reached and the people put aboard.

Mr. Root went out again, all of the crew of the Galveston believing that he would be drowned or killed. By this time the wind had risen so high that the whaleboat could not return with the twenty-one persons who had been rescued. Engineer Root had them placed in a large two-story building that was standing the strain better than others. The women and children of the party were weak from fright and hunger. No food could be found. Engineer Root made his way back to the store where he had obtained the rope and found a few crackers and other things, and he gave them to the party, with whom he remained until the storm was over.

In his second letter, dated September 11, Mr. Whitaker

tells of the terrors of the storm and its effects. He estimates the dead at 5000, and says that boatload after boatload of dead were then being carried to sea and put overboard. He reports the timely arrival of a steamer from New York having on board a distilling plant for the Galveston. "I shall set it up on deck and start it at once to make fresh water, for God only knows what we may need. People are coming all the time to the vessel begging for a drink of water and something to eat. We have had some people on board since the night of the storm."

He says that there are few families in Galveston that have not lost one or more members, and that in many cases entire families have been wiped out.

Mr. McIlhenny was rescued from the flood, but completely exhausted. He says the water came up so rapidly that he and his family sought safety upon the roof. He had his son Haven in his arms and the other children were strapped together. It was not long before a heavy piece of timber struck Haven, killing him. He then took up young Rice, and while he had him in his arms he was twice washed off the roof, and in this way young Rice was drowned.

Mrs. Lucy's oldest child was next killed by a piece of timber, and the younger one was drowned, and next Mrs. Lucy was washed off and drowned. Finally the roof blew off the house, and as it fell into the water it was broken in twain, Mrs. McIlhenny remaining on one half and Mr. McIlhenny on the other. The portion of the roof to which Mrs. McIlhenny clung turned over, and this was the last seen of her.

Joseph E. Jonson, a prominent citizen of Austin, who was among the list of missing and has turned up, saw some horrible sights which he can never forget. He said:

"Many of the survivors got through the flood almost by a miracle. I saw young men who were black-haired on Saturday come from the ordeal with hair turned completely white on Sunday. One young man in particular who was rescued by myself and others had been floating around the surging waters all night on a plank, expecting every minute to be drowned or crushed under falling buildings. His hair has turned gray, and he was a physical wreck.

"It would take 5000 men one year to clear the streets and town of Galveston of the wreckage, so complete is the ruin. Words cannot describe the chaos that has been wrought. The biggest liar in America could not do justice to the existing condition of affairs there.

"I was in the Tremont Hotel during the storm. The building was thronged with refugees. Men were praying throughout the night, and above the roar of the wind could be heard the crash of buildings and the splash of the waves against the buildings.

"We expected the hotel to go down any minute. At daylight Sunday morning four others and myself started out to view the ruins. We passed eight bodies within a block, and when we reached the beach, where the waters were still running high, we stayed some time, and about one body per minute passed floating with the tide. Homes that formerly were elegant mansions are a mass of wreckage. In one such home we found an old man and his wife and daughter and servants gathered in the parlor. In the parlor with them was a cow and other animals which had sought refuge in the house. We got the cow out and did what we could for the sufferers. This is only one incident of many similar ones."

Among the refugees which the Galveston, Houston and Henderson train picked up at Lamarque, which is about four and one-half miles south of Virginia Point, was Pat Joyce, who resided in the west end of Galveston. Joyce told a harrowing tale of hardships he had suffered to reach the mainland and his experiences after he had left Galveston.

"It began raining in Galveston," said Joyce, "Saturday morning early. About 9 o'clock I left for home. I got there about 11 o'clock and found about three feet of water in the yard. It began to get worse and worse, the water getting higher and the wind stronger, until it was almost as bad as the Gulf itself. Finally the house was taken off its foundation and entirely demolished. People all around me were scurrying to and fro, endeavoring to find places of safety and making the air hideous with their cries.

There were nine families in the house, which was a large two-story frame, and of the fifty persons residing there myself and niece were the only ones who could get away. I managed to find a raft of driftwood or wreckage and got on it, going with the tide, I knew not where. The Gulf and bay were full of wreckage of every description, and it seems as if every frame house in the town must have been blown down, judging from the amount of driftwood that was floating about."

Dr. S. O. Young, secretary of the Galveston Cotton Exchange, was on the boat going out of Galveston.

"I am going up to Houston," he said, "to buy some clothes. These I have on are borrowed."

Dr. Young's experience in the storm was unique. His house, standing at Eighth street and Avenue M, was the last to go down in that territory at the southeast corner of the city, that was swept clean of human habitations. Thus he had a chance for his life, for already the wreckage of other buildings in his neighborhood had passed by and been heaped up in the huge swath of ruin on slightly higher ground.

"My family were up North," Dr. Young said, "and I was alone in the house. I was about to take a bath when the storm became so terrific, and so I was dressed for swimming. My house had a pretty firm foundation, and, like many others, I thought it would withstand the elements. It did stand longer than any of my neighbors' homes, and to that fact I probably owe my life.

"For hours I stayed within and watched the wrecked houses of my neighbors sweep by. I was preparing for the worst. My house creaked and swayed. A door jamb had been torn partly loose. I wrenched it off as I saw my house was going and swam out clear of the falling timbers."

"At this time practically all of the wreckage had been washed in toward the center of the city and the water was comparatively free of timbers, so I simply went with the waves, clinging to my plank, until I was thrown up on top of a pile of debris and got a firm footing. I was bruised, but able to look after myself. Think what it must have been for those who were in the midst of the house wreckage."

Colonel McCaleb is a man of iron, a soldier descended from a long line of soldiers, but his heart is tender. Sitting at table in the dining hall of the Tremont, his only resting time, he told some touching stories of his experiences at headquarters office during the first few frantic days. Here is one of them:

"Two or three days after the storm there came into the office a big fellow named Lacy, a cotton jammer. I had never seen him before. He had a voice like a foghorn. He had been working hard burying and burning bodies, digging into the drifts, ever since Sunday morning.

"But Lacy had a grievance, and grievances were not proper those days. He stamped and stormed. He was not being treated right. Lacy had lost everything in the wreck, and he had an idea that he was not getting as much relief as he was entitled to.

"I asked him how much he required for his family.

"'My family? They are all gone,' he replied. 'There's nobody left but me.'

"I had been a little angry at the man, but this touched me, for I had my wife. I began to tell him how I lost my home and everything I owned, thinking to console him by establishing a fellow-feeling. 'Why, Lacy,' I said to him, 'all I have left is this red shirt I am wearing, and these trousers and shoes. I haven't even a change of underclothes. I haven't got a cent, Lacy.'

"The great big cotton jammer's anger subsided at once. He put his hand in his pocket and drew out a dollar bill. Offering it to me, he said:

"'Colonel, I'm awful sorry for you. Here, take this. I've got one more left.'

"I reckon," concluded Colonel McCaleb, "that if I hadn't been a military man I should have cried."

Then McCaleb told a little tale in a different vein. He said that a day or two ago a woman came into headquarters office and asked to see him personally.

"Colonel," she said, "I want you to send a file of soldiers down to my street. Some folks around the corner have got my pug dog and they won't give it up. They say it belongs to them, but it doesn't, and now I want you to send soldiers out there and make them give my doggy back."

"Madam," said McCaleb, with dignity, "this city has not been put under martial law for the purpose of recovering pug dogs who have forgotten whom they belong to."

Previously Colonel McCaleb had told me that the loss of

pictures of his ancestors for several generations, swept away with everything else, had grieved him deeply, for he valued such relics highly.

As he took his old straw hat off the hatrack on leaving the dining hall he turned to me, and taking a battered little photograph from under the inside band, he said:

"The other day I had business out in the vicinity of where my home used to be, and I found a piece of my house. I recognized it by the boards, and in the mass of timbers I found this. It is a picture of my mother and myself, taken when I was a little boy. Money can't buy that from me."

Clarence Howth, writing to his brother in New York, speaks as follows of his experiences:

"All is gone! My poor, sick wife, her father and Mr. Belts, the only friend on earth, are drowned, and their bodies, with the countless other dead, without discrimination, are being piled together in carts and burned or buried at sea.

"This is the first day I have been able to walk since the storm. Our house was the last to go. We were all in the attic when the house gave way—Dr. Sawyer, Marie, a trained nurse and myself. The sea was even with the attic windows and the waves dashing over the house when the house gave way.

"We were all buried underneath the ruins in forty feet of turbulent water. The house went down a little after dark. The night was black with darkness, the wind blew at one hundred miles an hour, and the rain poured with intensest fury. It was intensely cold.

"I never heard a sound after the crash. Marie was in my arms, but the crash jarred us apart. I was fastened under the ruins, it seemed to me, an hour, and was strangling—almost the last death struggle—when I was released and made my way to the top.

"I struggled for life for ten hours. Am bruised and cut, but am being well cared for. Martial law, consternation, ruin and starvation prevail. My limbs are left and my mind only slightly impaired, and I can begin life again as I entered it."

Ashby Cooke, a young business man of Galveston, was engaged to marry Miss Addie Rogers, one of the fairest of Southern belles. Miss Rogers and her mother lived near the beach in the district where all the houses were washed away. Saturday evening Cooke waded and swam to the Rogers residence to look after the girl he loved. Part of the way he climbed along walls and fences, and where he could not climb he swam.

Reaching the house, he found Mrs. Rogers and her daughter terrified. The water was rapidly mounting into their house and they helpless. Cooke helped them pack their valuables in a little trunk, hoping to save these few belongings at any rate.

The three were upstairs, the water in the rooms below being deep. When they realized that the house must fall soon Cooke secured a rope and let the trunk down into the water outside. Then he let Mrs. Rogers and Addie down, telling them to use the trunk for a support until he could follow them.

Just then the top of the house crashed in, and Cooke was caught by the rope and choked to death against the window sill.

J. A. Fernandez, a citizen of Austin, went to Galveston a few days after the storm, and thus describes what he saw:

"In the course of my rounds I saw a family of six half naked. They appeared crazy, and would look into the face of every stranger with a vacant stare that was pitiable. They were hurrying in the direction of the places where provisions were being distributed. They had lost their home and had only what little clothing was on their backs. There were thousands in a similar condition.

"The soldiers are doing effective work in burning the dead. The bodies presented a horrible sight, and swollen to the size of a barrel. Limbs would in cases fall to pieces as the bodies were being thrown into the flames. Soldiers are working heroically, and do not shirk the arduous duty. There are at least five hundred people whose minds have become unbalanced, and some have lost every vestige of their mental faculties.

There are some raving maniacs among them, one of whom came under my personal observation. His name is Charles Thompson, a gardener. He occupied a room above me at the hotel, and during the night he kept raving and pacing the floor and kept calling on God to witness his actions, continually invoking the mercy of the Deity. He has lost his family and home, and by a miracle saved himself.

"As soon as he was out of personal danger on that awful night he commenced rescuing women and children, and saved 700 people, according to a gentleman who knew the circumstances. He then lost his mind. He created so much excitement at the hotel that two policemen were detailed to capture him. He heard them approaching and leaped out a three-story window to an adjoining building. His fall was somewhat broken, but his body struck against the window in my room. He was badly injured, but continued his mad flight. He baffled his pursuers and escaped. This occurred at 5 o'clock in the morning. This is only one illustration of the conditions that prevail.

"A man whose wife was drowned in the flood had been searching in vain for her remains for several days, and yesterday located the body in the water near Thirty-third street and Avenue O. Soldiers had also seen the body and they took it in charge. The husband protested and rushed wildly to take possession of the body. The soldiers had to discharge their duty, and the husband, practically demented, was bound while the body was thrown into the flames. The man made frantic efforts to get away from the soldiers, but to no avail."

———.

W. J. Johnston of Galveston, a Postal telegragh operator

ELEVATOR "B."

NEAR 18TH ST. WHARF.

who formerly worked in St. Louis, told a pathetic story of the storm while dining at the Tremont hotel in Galveston ten days thereafter.

"My house at Eleventh street and Avenue L was swept away," he said, "and my wife and our two babies were lost. I put them on the inverted roof of another house that was floating by. Then I climbed on and we floated for half an hour. The tempest was so furious that I was unable to make my wife hear my voice, though I shouted as loudly as I could and our faces were but a few inches apart. I lifted up the faces of my little ones to see if they were crying. They were not crying, but terribly frightened.

"A big breaker swept over us, the water full of timbers and other wreckage. All my folks were gone when I recovered from the shock of a plank's blow on my head.

"I went to work as soon as I could walk, and have been sending out press matter night and day since. I could not live if I did not work this way. I would go mad.

"I was told that my wife's wedding ring, with our initials and the date, 1894, was at the headquarters claim office. I went and got it. Father Kirwin had left it there. The priest told me that he took it from the finger of my wife, whose body had been placed on a barge with hundreds of others to be buried at sea.

"Father Kirwin had not known my wife. He said her body was the only one that could have been identified in the whole bargeload. It was not swollen or discolored. He had it laid aside and sent up town to try to get someone who could identify it, but after a few hours it was put back on the barge.

"Father Kirwin and a negro whom I know tied three window weights to the body to make sure that it would sink. The wedding ring is all I have left."

Mr. Johnston spoke apparently without emotion. Other men at the table were moved to tears.

"You don't look as if you have much sorrow," remarked one.

"No," he replied, "I don't look it, but—."

One of the terrors which tormented the survivors in Galveston was the fear of a recurrence of the wind and flood. One of the editors of the News thus describes his feelings while looking at an approaching thunder shower:

"As I begin the story at nightfall the lightning is illuminating the bank of clouds massed over the Gulf horizon. For the past half hour I have looked upon the flashes, and those around me wondered if it were to come again. The 'it,' of course, means the visitation of last Saturday night. They look anxiously around as the streaks of gold and silver illumine the sky at quick intervals. My friends are those who went through the awful experience of the cataclysm. I know as well as mortal man can know anything that this island is no longer a target for the elements. I know that a target like this devastated island could no longer invite the shafts of the elements, even if the elements were endowed with human or divine intelligence. And I know in the simple faith of humanity that the God who 'plants His footsteps in the sea and rides upon the storm' would reach out with His omnipotent arm and throttle the agencies of nature if they should again aggravate wind and wave to vent their wrath upon these desolate shores.

"I know that if the sorrows of this community, what remains of it, have thrilled humanity they must have touched the wellsprings of divine mercy and sympathy, and that the helpless victims who have survived the tragedy of this moment may feel safe from another attack from the remorselessness of the storm.

"Galveston, stricken and bleeding, is safe from the wrath of all powers, human or divine. The vivid lightnings may cleave the sleepless waves of the sea and the thunders may play at will among the fantastic clouds in the sky. Galveston, soothed and compassed by the tenderness of mankind, is veiled in the folds of heaven's mercy, and the shrieking tempest is now but a whisper from the sky, the angry wave but the gentle falling of tears from above the stars.

"It is so hard to write the story or a chapter of it without feeling the power that appals human intelligence, just as it is hard to disassociate overwhelming sorrow from that broad sympathy which we do not understand, but which never fails to nestle close to human misery. Call it what you may, it is part of human life, and its presence comes when disaster overwhelms to bring humanity in the presence of God."

D. B. Henderson, a prominent citizen, who acted as chairman of transportation after the storm, was constantly on duty at the wharf from which the little steamer Lawrence plied to Texas City laden with refugees. One night a party of newspaper men met Henderson as he was coming off duty and induced him to talk about the storm.

"Gentlemen," he said, "I have ceased to tell about the storm, but there is one thing that I may tell you, and that is the impression made upon me by the corpses that continue to float into the strip where I am engaged.

"Now, I don't mind seeing dead grown folks float in, but when I see the body of a child it just breaks me all up and I have to go away from the sight of it.

"In order to explain this feeling I suppose I'll have to tell you about my part in the storm. I wasn't any hero at all. I had no chance to be one, and there was the horrible thing about it.

"I was at home with my wife and five little children. The house rocked and swayed. The water rose higher and higher. My wife's face was like cold marble. My babies were huddled there in a corner of the room, too terrified to cry, numb and dumb with terror.

"I stood by and watched them. There was nothing else that I could do. My utter inability to do anything to save those I loved, that was the worst of it all.

"I tell you I was not afraid of dying myself. But I felt that every moment would be the last for my wife and children. Weaker than I, they would die first. I had no hope that the house would stand. With every shock of wind and wave I expected it to tumble down about us. And yet I could do nothing—I, a strong man, absolutely helpless in the fury of the elements.

"Now, I thought, they are going. They will be swept away and die, and for a little space I will be left alive. Let me tell you that the keenest agony I experienced that night was in contemplating my own suffering in that brief interval between their death and mine, those crouching, huddled children of my blood.

"That is why I shudder and nearly faint when I see a child's body float into the slip."

In Galveston there was a man so utterly bereft of wordly goods that he was willing to sell the tomb of his own child for money enough to get his living children and wife out of the city.

He was of political prominence, having held one of the highest offices in that part of the State. He was born and reared in Galveston, but he is determined to send his family to St. Louis or Chicago and follow himself when he can dispose of his city residence lots, denuded of houses, for a few dollars.

"My home and all it contained, all I had, was swept away," he said. "Last week I was worth $12,000. Today I have not ten cents to buy a breakfast.

"I have one residence lot which cost me $2500. I will gladly sell it for $250.

"Out there in the cemetery I have a child entombed that

died nine years ago. I built a splendid vault, costing $3000. My wife and children must be sent away from this place, and I am willing to bury my dead child's remains and sell the vault to raise a stake so that I can get my folks out of Galveston."

Later this man offered his residence lot for $50.

One happy man I saw in Galveston. That is the pleasantest memory of the place I have. It was an old negro garbed in a gunny sack. He had lost his cabin home down on the flats and came out of the flood perched high and dry on the roof of a house that still stood.

He was stark naked and nearly frozen, but the warm sunshine of that Sunday morning following the night of doom thawed him out, and the old man made for a darky shack to get his nakedness covered.

"Ah didn' know but what Ah mought git 'rested for indecent 'sposure," he told me, "so Ah mosied roun' foh some duds. De culled folks didn' have none; dey all done get toted off by de water; but Ah kim down yah on de w'arf an' dis yah soldier guv me dis yah gunny sack, an' it does pow'ful nice."

It was a long, large gunny sack, such as they use in which to bring cotton to the place of baling. The old negro had cut holes for his thighs through the closed end and holes for his arms through the top. The open end he had tied over his shoulders with strings, and thus he was rendered not amenable to the law.

He was walking near the wharf carrying a basket filled with supplies doled out to him at the ward relief station. He sang a plantation melody, "Massa's in de Cold, Cold Ground."

"Where do you live now, uncle," he was asked.

"Huh! huh! I'se libin' outdoahs now, sah. Dey didn' git none of mah folks; we's all safe an' soun'. We's libin' under a roof from some folk's house dat got cocked up on de side. Dis ole niggah's satisfied."

Capt. Victor N. Theriot, a hero of the Spanish war and of the Galveston disaster, told a story of mother-love.

"Along toward morning I found a woman with a baby in her arms about a month old. Both were stripped naked by the storm. The mother was clinging to a ledge on a wrecked wall, hugging her baby to her breast. She was nearly frozen.

"The water had partially receded, but the wind still blew with much violence, and it was a chill wind. I called another man to help. We got a garment to cover the woman's nakedness, and I told her to hand me the baby, so that I could carry it to a place of safety, while she was taken in charge by another man.

"The woman hesitated. She looked at me searchingly. I suppose I was not very handsome looking after being out all night in deep water.

" 'I guess I can trust you with it,' she said at last, 'but before I give it to you you must promise to hold its head away from the wind. I have sheltered it this way all night. The wind musn't be allowed to touch my baby.'

"I promised, and she handed over the tiny infant. By George! that sort of broke me all up."

In an emergency hospital at Houston in the Auditorium Building I saw a man with his right hand and left foot swathed in bandages. He was John Holman, a dairyman who had lived "down the island." His house was near the Catholic Orphan Asylum, which was utterly wrecked. Holman said his own house was carried away, and of a family of ten he was the only one left.

"The orphans were nearly all lost," he said. "Only three out of more than a hundred were saved. One of these was a little girl about three years old. They found her Sunday, sound asleep in a salt cedar tree, three miles from where the asylum had stood, with not a scratch on her.

"Two little boys were picked up alive. One was badly hurt. I carried him to a place of safety the day after the storm.

"I was alone in my house, all my folks having gone to another house which they thought was more secure. When my house went down I caught hold of a part of the wreck and clung to it. Part of the time I was underneath and part of the time on top.

"Finally I was able to swim for a cottonwood tree. I held on to the tree, twenty feet up, waist-deep in water, till morning. Then I crawled and swam into town.

"Lying across the railroad tracks I saw the body of a sister from the asylum. Her hands were extended in front of her and she clasped her little cross. About fifty corpses of the little orphans lay near her."

If there is one sight a little more frequent than all others in the streets of Galveston it is the section of tin roof, varying from a square yard to a dozen square yards. If there is a roof in the city which was not damaged it has not been reported. For tin roofs the hurricane seems to have had a special fondness. It rolled up the fragments and deposited them in queer places, sure enough. The houses of Galveston abound in porches or verandas or, as the popular name here is, "galleries." It was the ambition of the hurricane to leave at least one curled and battered section of tin roofing on each gallery in the city, and the ambition was nearly attained.

Perhaps the next most frequent object is the wandering cistern or tank. An indispensable adjunct to each house in Galveston was the wooden tank, which occupied a conspicuous position in the back yard. Custom decreed that the cistern should have a position elevated several feet above the ground. Into it the rain water from the house roof was led. This insured a domestic supply of soft water. The drinking water came from artesian wells on the mainland, and was conducted in mains some miles. The wind bowled over these wooden cisterns, and the rushing water carried them through the streets. In short, the tank was the sport of the hurricane.

CHAPTER VI.

SIGHTS IN GALVESTON.

STRANGE REMINDERS OF THE HURRICANE ABOUND IN THE WAVE-SWEPT CITY.

With the hurricane a many-told tale, Galveston still abounds in strange sights. Street cars traverse several of the former routes, but they are of the pattern discarded years ago, and the motive-power is the "hay-burner," which is the name the superintendent applies to the mule. There are tangles of wires hundreds of miles long to be straightened before it will be safe to turn on electric currents. Galveston had a full complement of telephone, electric-light, telegraph and trolley wires, all overhead. The hurricane played all manner of freakish havoc with poles and cross-bars and wires. It left not so much as a block of the network in place. It not only threw down poles, but it twisted and whirled into snarls the wires they bore, until restoration will mean the untangling and coiling of the down wire and the entire reconstruction of the lines in most parts of the city.

The mule car is of slow schedule, but it goes fast enough for the vision to take in the wonderful things to be seen.

In a house with the front gone a colored woman is busy with the "wash," so busy that she doesn't look up from the ironing-board as the car passes. The whole operation of the home laundry is on view to passers.

Some hundreds of homes which the wind and waves did not reduce to planks and small sections they moved and set down in most inconvenient locations. In the weeks to come, after the mountain ranges of debris have been disposed of, there will be a period of house-moving. A ride along Broadway lengthwise of the city shows every cross street leading toward the Gulf occupied by from two to twenty vagrant houses. Many of these buildings are in fairly good condition. They will have to be put upon rollers and hauled back to their original locations. In places these estrays completely fill streets so that there is no way past. In not a few cases the occupants of these houses have not abandoned them, but are using the street for the door-yards. In the swirling currents and the shifting winds these houses which floated off their underpinnings seemed to have moved by no general rule of direction. Some went one way and some went others, so that a man now finds himself with a neighbor on his left who lived next door on the right a few days ago.

These street locations, however, do not mix things so badly as where houses have been moved into lots where they do not belong. Here, for instance, is a resident who believed in plenty of ground space. He finds that his neighbor on each side of him has moved in on his spacious lot, while the cottage on the opposite side of the alley has floated into the back yard and located, for the time being, close up to the kitchen. What adds to the aggravation is that the traveling neighbors have not taken any pains to align their houses with the building which was there, but stand at independent angles to each other and to the street. The sight is enough to give one with devotion to orderly array a case of architectural horrors. Neighborhoods are sadly mixed up by this indiscriminate redistribution of homes. The only consolation is that the homes were left standing.

Tanks are everywhere in Galveston save where they belong. They are especially noticeable in the wide boulevard, Broadway, and in the small parks. Many of them were left on their sides, and furnished the first roof which some of the homeless refugees found to cover themselves Sunday morning. The tramp tank is still a familiar sight in the public places of Galveston.

Her live oaks were the glory of Galveston. They fringed the streets and made beautiful the spacious grounds of mansions. Through these trees the gale tore until many of the largest were left prostrate, with roofs on one side thrust up into the air. When the surviving residents came gradually from the dazed feeling which followed the storm they grieved for the lost live oaks. Then was witnessed one of the extraordinary spectacles. The suggestion was made that these trees, if they were lifted in place and their roots put back in the soil, might be saved. Some people gave attention to the salvation of the live oaks before they put their houses in order. On Tremont street one day a crowd gathered to see a citizen, with half a score of strong men, ropes and tackle, work a couple of hours at righting and replanting a fine live oak. The house of the citizen was still in a state of partial dismantlement, and bed clothing was spread over holes in the roof to protect the contents from the rain. In the eyes of the citizen that live oak was of first importance.

Cows graze on the grass sward which adorns the center of Broadway. Horses are stabled in a Tremont street store which has lost one side. Brute instinct was in some cases equal to human intelligence. Martin Mayo, one of the dairymen of the city, had sixteen finely-bred Jerseys in a new stable. The cows were put in their stalls when the weather became threatening. The water rose until it invaded the dairyman's house. It rose four feet on the first floor and drove the family upstairs. Out in the street the depth was nine feet. William H. Honneus, an old Massuchusetts soldier, who went from Boston to New Orleans as Ben Butler's orderly, and remained in the South, was with the dairyman's family. He had been employed in building the new stable for the cows.

"The floor of the house," said the old soldier, "was three feet above the ground under it, and that was two feet higher than the street. After the flod rose to the level of the sills I cut holes in the floor to let the water in freely, so that the building would not be lifted off its foundations. Then I watched and measured the rise. The family had gone to the second floor. Every few minutes I took a light and went downstairs and tried my stick. In an hour and a-half the water had risen up three feet. At four feet the rise seemed to slacken. When it appeared to me that there was a fall of an inch I said nothing. I wanted to be certain. I waited fifteen minutes and tried my measure again. There was a fall of two inches. I danced on the floor and called to the folks above that we were safe. The next measurement showed a decrease of four inches, and the one after that gave a fall of fourteen inches. From that time the water went out rapidly. There had been at least six feet of water in the stable. We supposed, of course, all of the cows were drowned, but after a time a 'mooing' was heard. Those cows had had the sense to climb with their forefeet in the mangers and feed-boxes, and standing on their hind legs, had been able to keep their noses above the water level. Thirteen of the sixteen were alive. Those that drowned were too short-legged for the depth."

The action of the dairyman was characteristic of the community spirit. He drove the thirteen surviving cows to the police headquarters and said:

"I haven't got the nerve to go about selling milk at fifteen cents a pint at such a time as this, and I can't afford to buy feed and sell for less. Take these cows and use the milk for the injured and sick in the hospitals."

Mayo came out better than most of the Galveston dairymen. One of them saved five cows out of forty and four horses out of nine. In the Galveston papers are appearing even now such reminders of the storm's confusion as these:

LOST—During the storm, some sixty head dairy cows and calves, principal brand A, with diamond center, on left and right hip. Ten dollars reward for every one returned. B. F. Mott, Broadway, between Fortieth and Forty-first streets.

LOST—During the storm, some 150 head of dairy milch cows, branded C. J. on right hip and a compassed S on left hip. Ten dollars per head reward for anyone finding the same. Juneman Bros., Broadway, between Fortieth and Forty-first streets.

Some of the well-to-do families kept their own Jersey cows grazing in lots near their homes. In several instances the cows were taken into mansions and shared the protection which the family had. At the home of J. K. Wiley, one of the finest in the city, resting on a Texas granite foundation, some 200 people found a safe refuge. When the water began to reach an alarming depth the Wiley boys remembered the pony and the Jersey cow and calf. They swam out to the stable and brought the pets into the house. Pony and cow were saved, but the little calf went under. Mr. Wiley had passed through the flood of 1875, and when he built his present home he made the foundations strong with granite, and for the superstructure employed pressed brick. The house passed through the storm with less injury than almost any other in the city.

In the long list of the dead the family name of Labett appears several times. Only a year ago five generations of the

Labetts lived in Galveston, but the family name was almost wiped out by the flood and storm.

A young man connected with one of the railroads was down town and escaped. When the parties of searchers were organized and proceeded to various parts of the city one of them came upon this young Labett near the ruins of his home and all alone. He had made his way there and found the bodies of his father and mother and other relatives. He had carried the dead to a drift of sand, and there, without a tool, with his bare hands and a piece of board, he was trying to scrape out gravel to bury the bodies.

When the waters swept up the streets and panic began to spread three policemen found their way to the city hall and reported to their chief, Captain Ketchnim, for duty. Hour after hour they made life-saving trips in various directions as the chief gave his orders. Houses were falling, the wind carried through the air bits of slate and glass. The officers piloted women and children through the water to the city hall. They rescued from the water many who would have perished before morning. Without one word of complaint of fatigue, their chief says, these men continued their life-saving all night.

Toward morning there was a marked subsidence of water and a lull in the wind. Then after 400 people had been brought to the city hall, and after many others had been quieted and assured of safety, the three officers came to the chief for permission to go and see how their own families had fared. It was then daylight. Two of them, Byrd and

Copyrighted by Leslie's Weekly, 1900. *Drawn by Gordon H. Grant.*
SEEKING THE WOUNDED AND SECURING THE DEAD AMID THE RUINS ON MARKET STREET.

Rownan, came back in a little while. They told the chief that their houses had been swept away and their families were lost.

In all great catastrophes I have yet to know of one that some special act of selfishness and brutality did not occur. There is hardly a great wreck recorded in which is not depicted the brute who pushed women from boats or from spars. In all I have heard of the thousands of incidents connected with this storm not one instance of that selfishness which would cause one person to deprive another of his means of escape has occurred. Thousands of instances of devotion of husband to wife, of wife to husband, of child to parent and parent to child can be mentioned. One poor woman with her child and her father was cast out into the raging waters. They were separated. Both were in drift and both believed they went out in the Gulf and returned. The mother was finally cast upon the drift, and there she was pounded by the waves and debris until she pulled into a house against which the drift had lodged. During all that frightful ride she held to her eight-months-old babe, and when she was on the drift pile she lay upon her infant and covered it with her body, that it might escape the blows of the planks. She came out of the ordeal cut and maimed, but the infant had not a scratch. Another man took his wife from one house to another by swimming until he had occupied three. Each fell in its turn, and then he took to the waves. They were separated and, as the persons above mentioned believed, they were carried to sea. Strange to say, after three hours in the water he heard her call, and finally rescued her.

But it is not necessary to go and recite these instances, for there were thousands, each showing that in time of danger, at least, the best sentiments in man's nature are aroused. It can be safely guessed that one-half of those who perished died in their efforts to aid others. The trite expression of "man's inhumanity to man" has no place in all that may be written or spoken of this great tragedy.

Mrs. A. Bergman, wife of the manager of the Houston Theater, was at a cottage on Galveston Beach. As the water rose she and her sister put on bathing suits. They waded and swam for a mile, when they met a negro with a high dray. They clung to the dray, but soon it was swept away and the negro was drowned. They began swimming again, but narrowly escaped death from flying roofs and wreckage before they reached a place of safety. They went to Houston, still in their bathing suits.

The residence of Branch Masterton, which was near Wholman's Lake, was carried clear across Galveston bay almost intact. Masterton and two children clung to the roof and landed near Texas City. Mrs. Masterton slipped off and was lost. Masterton dragged his house up on the beach and is living in it.

Capt. W. C. Rafferty, Battery O, First Artillery, abandoned his home and, taking his family, fled to shelter beneath the big ten-inch gun carriage at Fort Crockett. The water boomed in a deluge over the gun, but it kept on its foundation, and the captain and his family were saved.

A family by the name of Stubbs lived in the West End. The house collapsed. Stubbs and his wife and two children floated away on the roof. The roof broke up, and Mrs. Stubbs, with one of the children, was parted from her husband. The other child, five years old, fell off the roof, and it was supposed was drowned. When the flood subsided husband, wife and one child were reunited. Next day a soldier brought in the five-year-old child. The child had clung to a table until rescued.

On almost every hand one hears of heroic deeds during the storm, of people saved at great risk and others lost in endeavors to save others.

Mr. Milton, a well-known street-car conductor, is laid up as the result of his successful efforts to save three ladies.

Mr. Johnson, corner Avenue M and Twelfth street, remained in the house until it was down, and then attempted to save his wife, but she was swept out of sight, and his hand caught in the dress of another woman, whom he attempted to aid, but just at that moment he was struck across the face and stunned so as to cause him to lose his hold, and before he fairly recovered from the blow the woman was swept away.

In the West End Mr. John Sweigel placed his mother on a table in the house, while he and his four sisters remained in the room with her, as also did a nephew of Mr. Sweigel, ten or twelve years of age. The waters came up and washed them all out of the house. The boy climbed to the roof of the house, where he remained all night and was rescued next day. The mother and Mr. Sweigel and three sisters were drowned. One of the sisters was found clinging to the transom of the door and rescued Sunday morning.

Mr. Vianna, keeping a grocery on the corner of Avenue M and Twelfth street, floated on the roof on his house all night Saturday, but was rescued Sunday.

Capt. and Mrs. William Scrimgerm of Galveston, Texas, who have been visiting Captain Scrimgerm's sister, Mrs. C. C. Grafflin of 1006 Mosher street, Baltimore, have left for their home in Galveston. They received a telegram stating that their family was safe, but that much damage had been done to property owned by them.

John Rutter, aged twelve, was picked up beside a big trunk at Hitchcock, twenty miles from Galveston. The family consisted of parents and six children. The house collapsed, and all were lost, except the boy.

"I came up beside the trunk," said he, "and caught hold of a handle. Then I got on top and floated all night. Sometimes the trunk would lurch and I would be thrown into the water, but I hung on."

Mr. Otto Schoenrich of Porto Rico writes to his parents, in Washington, Professor and Mrs. Schoenrich of this city, that he passed through the storm on his way to Porto Rico by ship, but no damage was done.

Miss Celeste Heider of the Census Bureau, whose father, mother and two brothers were residing in Galveston, received a telegram from Charles Heider, one of her brothers, announcing that the members of the family were all safe, but had lost all their property.

Mr. Claude J. Neis of the Congressional Library also received a dispatch today from his brother, John J. Neis, announcing the safety of their parents and sister. The telegram closed, "Everything lost; thousands dead."

A report is current at Maplewood, a fashionable suburb of St. Louis, that Ernest Furniss, who left that place Tuesday evening for Galveston, after receiving word that his father and mother and three sisters had perished in the great hurricane, found that the information was correct and committed suicide.

There is another happy thing in Galveston. It is a little canary, homeless and orphaned of its mistress, but it sings and sings as it sits perched in its little cage hung in front of the door of a stranger who found it in the wreck.

Two days after the storm a citizen named Labotte was passing by a high heap of tangled timbers, household goods— and corpses. Suddenly he stopped and inclined his ear to listen.

From the midst of the mass came the sound of a bird trilling gaily. It was melody from a tomb. Labotte was amazed. He approached nearer and put his ear down at a gap between the timbers. The bird song increased in energetic rhythm.

Labotte went rapidly to a point where a gang of men worked at the debris.

"There's a live bird in this wreck," he said.

The men followed him, and with saws and axes and ropes hacked and tugged away at the timbers, until, ten feet down, they came upon a little cage, wherein a tiny, bedraggled, frightened canary hopped about.

Labotte took the bird home, where, recovered from the fright of having a gang of men chop it out of the wreck, it sings from morn to night.

And the beautiful belle or happy matron who loved and petted the bird in some other home—where is she?

PART III.

SKETCHES OF GALVESTON AND TEXAS.

A SKETCH OF GALVESTON.

Galveston before the recent flood which swept away its dwellings, destroyed its industries and carried to untimely death one-fourth of its population was the first city of Texas commercially and third in point of population. It is the county seat of Galveston county, and is situated on the northeastern end of Galveston Island at the mouth of Galveston bay. It was the largest port on the Gulf coast, with the single exception of New Orleans, which is really a Mississippi river port, and is 180 miles southeast of Austin and about fifty miles from Houston.

The city stands on such a slight elevation above sea level that it seems always destined to be particularly susceptible to floods and tidal waves driven upon her by those terrible tornadoes which have their origin in the region of the West Indies. This sandy stretch extends east and west for twenty-seven miles, and seven miles is its greatest width north and south. No part of the city is more than ten feet above the sea level. Where the city is built the island is only one and a-quarter miles wide. The island is intersected by many small bayous and bordered through its whole length on the ocean side by a smooth, hard beach, forming a fine drive.

The bay is irregular, branching out into various arms. It has a total area of 450 square miles. Its entrance is guarded by a bar, through which the present deep-water channel is constructed. The stone jetties flanking it are five miles long.

The city proper, being on the inner side of the island, is naturally protected from the sweep of the ocean storms, but the level of the island is so little above that of the bay that inundations have more than once been threatened. The bay is shallow in most parts, and the railroads reach the city on a two-mile trestle from the mainland.

Many years ago, before the settlers on this island presumed to claim that it would become not only the important seaport, but the great commercial and manufacturing center it was before the recent storm, Galveston, known all over the world as the beautiful and prosperous island city, had five rivals—Sabine Pass, Aransas Pass, Port Lavaca, Corpus Christi and Indianola—each in its way threatening to swallow it up, wipe it out of existence, as it were, the then little village on the Gulf of Mexico.

The remembrance of the struggle made by these several localities for supremacy is still fresh in the memory of many of Galveston's citizens, but the island home, appreciating the fact that nature had provided for her with a lavish hand, locating her where and how she is out in the great waters that have for ages, and will for ages yet to come, break against the shores of Texas, Florida, Alabama, Mississippi, Louisiana, the republic of Mexico, Central and South America, still lives and is the magnet that has attracted and with patient hand absorbed all its rivals, and instead of becoming the lonely and deserted "sand hill," as was predicted, is today the central figure on the Gulf coast.

Before the storm it was a city of nearly 80,000 prosperous

people, the jewel that shines brightest in the Lone Star State, and the gem that will yet sparkle in American history as the Union's most valued exporting and importing port, while Sabine Pass, Aransas Pass, Port Lavaca and Corpus Christi are still resting in the cradle of infancy, and Indianola is lost in the waters that made and are still making and not destroying the acreage of the "Island City."

Here was a city where millions of money were spent annually, a city on which the sun shines all the year round, where winter in its severest aspect is never known and autumn and springtime meet at the threshold of the year, where roses bloom in the gardens and scent the Southern breezes ten months out of the twelve, where figs and grapes and tropical fruits return a double fruitage, where birds never cease their songs, where everyone finds sweetest comfort in the blessings of today and the waters flow unfettered by the ice king.

The first census of the population of Galveston was taken in 1850, and the city then had 4177; in 1860 it had increased to 7307. During the civil war the population was much reduced, but in 1870 it had risen again to 13,800. In 1880 it was 22,253, and in 1890 it had risen to 29,084, and the census of 1900 gives it as 37,789. The growth of the city in point of population has been gradual, almost in exact proportion to its increase in trade and industry.

It may be said that the improvement of the harbor by the United States Government, whereby a free channel over the bar four hundred feet wide and about twenty-seven feet deep was obtained, was the factor which enabled Galveston to distance its competitors in the race for commercial supremacy in the Southwest. Somehow the people of the Island City had the energy to interest the powers that be in Washington to the appropriation of about $6,000,000 for the inauguration of a jetty system of harbor improvement. With a little advantage in the way of position and a good deal of hustle, the leading men of Galveston addressed themselves to the problem of getting a deep-water channel over the bar to the Gulf, and succeeded.

The Secretary of War was authorized in 1890 to let a contract for the work of building the jetties and dredging out the channel where it might be necessary. The law required that a channel of thirty feet in depth should be constructed, and the limit of the cost was to be $6,200,000. For the benefit of those who do not understand the principle of the jetty system of harbor improvement, it may be stated that it is an engineering scheme in which the forces of nature are harnessed to the task of scouping out and keeping clear the channel which it is proposed to construct. For this purpose a large bay connected with the open sea is necessary. Then a series of walls are made in the waters of the outer harbor in the shape of the letter V, the small part of the angle extending seaward, in which is a small narrow opening, at Galveston 400 feet wide. The arms of the angle extend usually to the shores, and are made of a great quantity of loose stone thrown into the water and coming near to the surface at high tide. The tide then flows in over these walls and fills the bay within. When the tide goes out the water in the bay has no direct outlet, except through the narrow opening at the angle of the walls. The water in this way scours off the bottom and creates a constantly deepening channel. As the current cuts down below the walls of stone recourse is had to revetments to prevent the edges from wearing away. In Galveston recourse was had to the dredge to deepen the channel in the first place, but the water has been constantly deepened from eighteen to twenty-seven and one-half feet and to twenty-nine feet by the action of the tides alone. The same system is in successful operation at Charleston, S. C.

Since the improvement of Galveston harbor by the United States Government it has become one of the chief ports of the United States, an immense amount of export trade which formerly came East having been deflected to that point. The harbor improvement made by the Government consisted of the construction of a system of jetties and the dredging of Texas City channel at a total cost of $8,700,000. By means of these jetties a depth of twenty-nine feet of water was obtained at Galveston, thus permitting of the largest size ocean-going vessels loading and unloading there. The city is connected with the entire railway systems of the United States and Mexico, and has a dozen direct lines of steamship communication with New York, Morgan City, the coast ports of the State and with Liverpool, Bremen and other ports. The trade of Galveston, import and export, is in the neighborhood of $100,000,000 a year.

There are three trunk lines of railway running into Galveston, the Missouri Pacific, the Santa Fe and the International lines. These before the storm crossed the bay on steel trestle bridges, each about two miles in length. These, together with the wagon bridge, probably the longest in the world, were all destroyed by the flood, a big steamboat being moored where one of them stood.

Attention was first directed to Galveston as a great seaport by its increasing exports of cotton. The State of Texas is one of the largest cotton-producing States in the South, and Galveston was the nearest port. By degrees it became the shipping point of a large part of the cotton grown in Southern Texas. The railroad companies were quick to see the advantages of its geographical position, so that the deepening of the harbor channel, the extension of cotton-raising in the State and the building of the railroads to Galveston were great enterprises which grew up together. And it was these considerations which prompted the people of the deluged town to take courage and to set about rebuilding.

In 1899 Galveston had become the first cotton-exporting point in the country, handling 67.91 per cent. of the crops. The city ranks fifth among the 127 foreign exporting points of the United States. In 1889 Galveston exported 6.39 per cent. of the merchandise that left this country. Its exports increased $10,500,000, or 13 per cent., in that year. The imports have shown a greater variety in the last few years, owing to the fact that the demand for import goods through Galveston is from a greatly widened territory. Galveston has recently become a great lumber port. The export of lumber and wood manufactures for 1898-99 amounted to $1,247,914, this business having been built up almost entirely since deep water was secured. The exportation of hog and dairy products, eggs and poultry is also a new business which has developed rapidly.

On the bay, on the north side of the city, is the commercial section, with wharves stretched for nearly two miles, lined with sheds and large storage-houses. In that portion of Galveston there were three elevators, one of 1,500,000 bushels capacity, one of 1,000,000 bushels, and the third of 750,000 bushels. The island from the north side was connected with the mainland by railroad bridges and probably the longest wagon bridge in the world, nearly two miles. In the extreme eastern end of the city there were many "raised cottages" built on piling and standing from eight to ten feet from the ground, as a precaution against floods, it being possible for the water to sweep under them.

The south side of the city, beginning within fifty yards of the medium Gulf tide, is the best residential part. One home cost the owner more than $1,000,000. Most of the houses, however, are of frame, but there are many of stone and brick. The streets are wide and straight, and the residence quarters abound in gardens shaded with magnolias and oleanders. Among the principal buldings are, or were, the new customhouse and postoffice, the cotton exchange, the courthouse, the Ball Free School, the Free Public Library and the Roman Catholic University of St. Mary.

The city has gas and electric-light plants, a water-works system valued at $450,000 and supplied from artesian wells, and numerous hotels, including one of the largest in the South, on the beach. In 1893 the gross city valuations were $25,000,000, the debt was $1,750,000, the city had authority to issue additional bonds to the amount of $1,500,000 for permanent improvements, and it owned property to the value of $1,955,560. In addition to being the great port of the South, Galveston is also an important manufacturing center. There are cotton compresses and cotton mills there which represent several millions investment. In the last few years the United States Government has spent $932,000 in the construction of coast fortifications at Galveston and the equipment of them.

There are two batteries at Galveston, one on each side of the deep channel, which the engineers have been constructing at that place. The channel is made by the jetty system, two jetties having been constructed across the bar out into the ocean. This channel runs in a northeasterly direction from the mainland and passes the island upon which Galveston is situated, the city lying south of the channel. There is a military reservation on the northwest side of the channel, and also another on the southeast or city side. It is probable that the great body of water came in from the ocean and up the channel, in which it would no doubt do a great deal of damage to the forts.

The fortifications are built of concrete on the sand, and could easily be seriously damaged by such a storm. The

Copyrighted by Leslie's Weekly, 1900.   *Drawn by F. Cresson Schell.*

1. THE SPLENDID CITY OF GALVESTON IN THE WILD FURY OF THE HURRICANE. 2. WRECKS OF SHIPPING AND GRAIN ELEVATOR AT THE GALVESTON WHARF. 3. THE GALVESTON STRAND, A PRINCIPAL STREET, IN THE HEIGHT OF THE STORM. 4. GALVESTON CUT OFF FROM RAILROAD COMMUNICATION BY WIND AND WAVE.

GALVESTON AND TEXAS COAST SWEPT BY AN AWFUL HURRICANE.

foundations are, of course, carefully prepared for the guns, but there never has been funds available for constructing breakwaters and protection against an inundation. There are in the fortifications at Galveston eight and ten-inch modern high-power guns, fifteen-pounder rapid-fire guns, 4.7-inch rapid-fire guns, six-pounders, twelve-inch breech-loading mortars.

The city of Galveston contains about 5000 acres, and is laid off in squares 300x260 feet. The streets are wide and intersect each other at right angles, and their general directions are with the points of the compass. Many of the streets have been paved within the past few years, and the street-car system of the city, connecting the business and residence sections of the town with the various beach resorts, was one of the best in the country. The buildings in the business section were large and substantial, and withstood the ravages of the hurricane well. While many of the cheaper structures near the beach were little more than shanties, many streets were lined with houses that would compare favorably with those in other cities. It is thought that the objectionable kind of building will be done away with in the future and that the new and greater Galveston will be a city that will not only be substantial, but will be of a solidity to laugh at the storm king hereafter.

Much attention was given by the people of Galveston to shrubbery and trees on the sidewalks and in the yards of the householders who could afford them. Flower and plants grow luxuriantly in the climate of Galveston, and it was a favorite custom of the people to cultivate them.

Thousands of pleasure-seekers from all parts of the South have visited Galveston in summer to enjoy its unrivalled bathing privileges. No other beach, perhaps, in the United States is better calculated for bathing than that upon which the great Pagoda and other famous bathing-houses stood a few short weeks ago.

There were thirty-five churches in Galveston before the flood, twelve of which belonged to the colored people. St. Mary's Cathedral was one of the most imposing of the church edifices, though there were many more that were close up to it in point of architectural attractiveness. After the storm the first Sunday found the most of the churches destroyed or so injured that they could not be used for religious purposes. The congregations were scattered, and there was almost no place for the worshippers of God to meet together. The Hebrews had an imposing synagogue, and the Methodists and Presbyterians some excellent churches.

The public school system was the best of any in the State of Texas, and the school buildings were comfortable and good houses. St. Mary's University, a Catholic institution, furnished higher education for many of the young men of the city, while the school kept by the sisters of the Ursuline Convent afforded similar privileges for the young women.

Galveston had about thirty hotels, and several of them enjoyed a very wide reputation among the people who travel as well-appointed and comfortable places at which to stop.

There are four national banks and a considerable number of private banking establishments in Galveston, having an aggregate capital investment of $10,000,000. There were before the disaster two daily and several weekly papers. Among the industries located there were the city gas works, the Brush electric-light plant, a cottonseed-oil mill, a large ice factory, four sash and blind factories, two flouring mills, steel and brass foundries, eight cotton compresses and a long list of minor mills and shops.

The harbor of Galveston is on the north side of the island fronting the bay. Here the railroad bridges centered. Here was the Grand Central Depot and the freight-yards of the several lines. The wharves extended along this waterfront for more than two miles, on which were the great sheds for the storage of cotton and the wheat elevators, with a capacity of 2,500,000 bushels of wheat and corn. This harbor with its wharfage was the best on the Gulf coast and one of the best in the United States. It has natural capacity for a vast commerce, which "manifest destiny" has brought to it in years past and will continue to spread upon its wharves in years to come as the resources of the Southwest are developed and flow to Galveston for an outlet to the markets of the world. At the entrance to the bay from the Gulf lies the crescent-shaped bar of sand which the waves of the sea have piled for ages across the outlet. It was through this that the army engineers have cut the passage for the entrance of the largest ocean ships.

The recent hurricane is not the first, though it is by far

the most severe, disaster which has befallen the city. In 1872 the entire east end was swept away by a tidal wave that followed a terrific storm, which swept the Gulf coast for three days.

On the night of November 12, 1885, the city was visited by a destructive fire, which within five hours swept away forty-seven blocks of the most densely-populated portion of the city. The fire started in a steam planing mill on the dock. A stiff north wind was blowing at the rate of forty miles an hour, and the flames quickly gained such headway that their force was spent only when the last building in its path had been consumed on the opposite side of the island. The area of the burnt district was about 100 acres, and upwards of 500 buildings were destroyed. The property burned was valued at $2,000,000, the insurance on which was $1,200,000. This sum was promptly paid by the companies, which enabled the burnt district to be rebuilt by a superior class of houses, greatly improving the city.

Large sums have been spent by the Galveston Wharf Company in the improvement of the internal facilities at Galveston. Owing to the competition in the export trade of the Southwest, it became necessary several years ago to reduce the costs of the transfer of freight from the railroads to ocean ships to the minimum. Accordingly, the rates of handling cotton, wheat, oats, rye, corn, hay, cattle and hides have been placed at the lowest figure, and this fact, taken in connection with admirable transportation facilities centering in Galveston, made it the chief emporium of the Southwest

The transportation facilities remain, and the wharves can be repaired at much less cost than a new trade center could be established elsewhere. This accounts for the energy displayed by the railroad companies and others in rebuilding Galveston and seeking to make it a larger and better city than it was before the visitation of the West India hurricane, which laid its proud supremacy low and left the puny works of man the jest of storm and tide.

It used to be the custom of the Galveston papers to record with scareheads on the first page the arrival and departure of tramp steamers which entered over the bar to load with cotton and wheat for foreign ports. Back in 1891 this was one of the diversions of journalism in the Island City. On August 29 of that year six steamers were in port at the same time loading with wheat and cotton for foreign ports, and the local papers recorded the fact as one of the most significant in the history of the port, and the local press indulged at that time in dreams of the coming greatness of Galveston, none of which reached in their apparent extravagance the truth of ten years later.

The history of Galveston may be divided into four periods—first, the days of the filibusterers and buccaneers from 1816 to 1820, followed by the practical depopulation of the island for about ten years. Herrera, the minister of the Mexican patriots, learned of the security of Galveston harbor and decided that he would take possession of it. This adventurer accordingly sailed to the island and landed upon it September 1, 1816, having in his company Commodore

Aury in command of a small fleet of vesels of the republics of Mexico, Venezuela, La Plata and New Grenada. On September 12 of the same year a government was established, and Aury was chosen governor of Texas and Galveston Island, taking oath of allegiance to the existing Government of Mexico. Soon after this Aury's vessels were sent out to cruise along the Spanish Main, and it soon swept the whole Gulf of the shipping of Spain.

About this time Xavier Mina, who had distinguished himself in the peninsular war with France, entered into an agreement with Toledo to wrest Florida from Spain, but when Toledo went over to the King of Spain Mina sailed to Galveston to co-operate with Aury and Perry, landing there November 26, 1816. Aury's cruisers made Galveston prosperous for a time. Slavers were brought in, but as there was no market for slaves in Texas at that time, they sought a market for them in the United States, receiving good profits for their piratical undertakings. Companies were formed in Louisiana, who received the slaves either at Sabine, Point Bolivar or Galveston, took them to a customhouse officer in Louisiana, denounced them as imported slaves, had them sold under the law by the marshal, and repurchased them, pocketing half the money as informers.

The forces in Galveston at this time consisted of about 350 men under Aury, besides his fleet, and there were 200 men under Mina, while Colonel Perry had 100 men at Bolivar Point on the mainland. They sailed or marched to take part in the war of Mexico against Spain. A few days after Aury and his men had sailed from Galveston Lafitte and his buccaneers arrived. By the end of 1817 Lafitte's followers in Galveston numbered about 1000 men. The town built by Lafitte on the ruins of that of Aury was called Campeachy.

About this time General Lallemand, an artillery officer under Napoleon, came to Galveston with about 100 Frenchmen. Proceeding a short distance up the Trinity, he built a fort, intending to cultivate the soil, but was induced to abandon the enterprise, owing to the opposition of the Spaniards. His followers returned to Galveston, and a part of them remained with Lafitte. The buccaneers were gradually driven off by a United States naval vessel. The floating population of the island during the days of the filibusterers varied from 50 to 1000.

The project of establishing a city on the east end of Galveston Island took shape in 1837 and began to be carried into effect in the spring of 1838. Prior to the battle of San Jacinto, and after the expulsion of the Mexican authorities in 1835, Galveston had regular commerce, although vessels frequently arrived with immigrants and supplies as an aid to the Texas revolution. In 1835 a military post was established in Texas and maintained until after the expulsion of the Mexicans in 1836.

The present site of the city was acquired from the republic of Texas under special grant by M. B. Menard and associates, and its first incorporation as a municipality was in March, 1839. Its population at that time was about 1000. The settlements in this portion of Texas were confined to a few counties contiguous to the coast, and the commerce was correspondingly limited. The foreign trade began in February, 1840, when a French vessel arrived with a cargo, and in the following winter British and German vessels followed. After Texas was annexed to the United States the commerce of Galveston steadily increased.

The secession of Texas led to the blockade of the port of Galveston during the civil war. It was captured by the United States forces October 8, 1862, but was regained by the Confederates January 1, 1863. Since the war Galveston has steadily grown from a small town to the largest seaport in Texas.

A brief study of the map will show why Galveston, built on a sand bank not fifteen feet above the level of the sea, has become the great emporium of the Southwest. Its location invited the railroads, and they and the terminal companies have done the rest. It would be as impossible to move the chief seaport of Southern Texas from Galveston Island as it would be to place Chicago, Cleveland, New York, Buffalo, Boston or San Francisco somewhere else, should disaster overtake them as it has wave-smitten Galveston.

A SKETCH OF TEXAS.

Texas is the extreme Southwestern State of the Union, the Rio Grande separating it from Mexico and the meridian 103 forming the line between it and New Mexico. On the north the boundary is marked by the Red river from Louisiana to the 100th meridian, thence northward by that meridian to its intersection with the parallel of 36° 30' north latitude, and thence to the meridian 103. The State has 265,780 square miles of territory, being 760 miles in breadth and 620 miles long. The geographer of the Census Bureau has estimated the water area of the gulfs and bayous along the Southern coast and the lakes and rivers at 3490 square miles, leaving 262,290 square miles of land. There are twenty-five counties in the State, a few of which, in the Western part, are not yet much settled and are poorly organized. These counties vary in size from 150 square miles to 25,000 square miles, being larger than several of the richest and most powerful States.

Texas is one-half as big as Alaska. Connecticut, Delaware, Maryland, Vermont, New Hampshire and New Jersey are about the size of as many counties in the Lone Star State, while the State of West Virginia has a little less area than the largest county of Texas. The States of New York, North Carolina, Utah, Nebraska and Rhode Island could be lumped together, and their combined area would lack 430 square miles of being equal to that of Texas. If the State of Texas could be laid down in the Eastern part of the country it would cover all New England, New York, Pennsylvania, New Jersey, Maryland, Virginia, West Virginia and one-half of Ohio. To travel the distance across the State in a direct line a man must take the train at Portland, Me., and journey to Richmond, Va., and in the other direction the journey would extend from New York to Cleveland. Texas is larger than either the Austrian or the German empires, France or the Islands of Great Britain.

The inhabitable portion of the State includes about 130,000 square miles, and the remaining area is composed of the Southwestern prairies and the gypsum lands, which for years to come probably will not be put to agricultural purposes.

In the State of Texas are to be found the greatest diversity of soil and topography, the former passing from the extreme fertility of the valley of the Red river on the north, the Brazos in the middle of the State, and of the Rio Grande on the south, to the extreme sterility in the sandy desertlike prairies of the Southwest. The topography passes from the flat and low coast country by gradual transitions to the chains and peaks of mountains in the Far West, whose summits are 5000 feet above the level of the sea. These extremes in soil and topography are matched by the diversity of the population. In the east and central portions of the State the counties are thickly settled, and there are large and flourishing towns and cities, while in the west and extreme southwestern parts few people live. To complete the picture of extremes, we find that several of the great agricultural belts of the South, that form so prominent a feature in the other Southern States, have their termini in Texas, and are cut off either by the prairies of the coast or by the mesquite and cactus chaparral prairies of the Rio Grande region, or they abut against the Eastern bluffs of the Western plateaus.

The coast of Texas presents features at variance with

those of any other State. In other States the coast may be cut up by large bays extending inland many miles, while it is here bordered by an almost continuous chain of islands and peninsulas. These islands are elevated but a few feet above the sea, and, like Galveston Island, are at times subject to inundation by storm and wave. The Gulf border of these islands is very regular, extending in almost an unbroken line from the mouth of the Sabine river to Corpus Christi, which occupies the highest point on the Texas coast, and then turns with a regular curve southward to Mexico. These islands, which are separated from the mainland by distances varying from two to twenty miles, are covered with heavy belts of sand and sand dunes, which rise fifteen or twenty feet above the beach. The latter skirt the shore line for many miles, and, as on Galveston Island, are usually broad, and offer many inducements to pleasure-seekers. The longest of these islands is Padre-Island, which extends from Corpus Christi bay to near the mouth of the Rio Grande, a distance of more than 100 miles. The large estuaries that have been formed at the mouths of the streams, except the Sabine, the Rio Grande and those of the Brazos region, form another feature peculiar to the Texas coast. The border islands of these estuaries are usually high, their almost vertical clay bluffs being washed by the waters of the bay, and the open prairies of the uplands often extend to their very edge.

The large territory occupied by the State naturally presents a variety of climate, and while the coast counties are warmed by the sea breezes during the winter months and have a mean temperature of 53° in December, the northern counties along the Red river suffer severe weather, the temperature of Denison for the same month being 41°. The minimum and the maximum extremes during December are at Galveston 18° and 72°, and at Denison 2° and 76°, as shown by the records of the Weather Bureau. At Corsicana, an intermediate point, the extremes are 6° and 80°, with a mean for December of 47.4°. Brownsville, situated more than three degrees south of Galveston, has for the same month a minimum of 18° and a maximum of 83°.

During the summer months the northern counties of the settled portion of the State enjoy cooler nights and have hotter days than those on the coast, though the mean temperature is the highest on the coast by several degrees. July at Galveston and August at Denison are the hottest months of the year, the average temperatures being, respectively, 83° and 80°, with maximums of 93° and 101°. Eagle Pass on the Rio Grande seems to be the hottest place in the State, its maximums for the months from the first of March to the last of July being greater than those recorded at any other point during the same period, and that for the months of June and July 108°, being the highest recorded in the State. At Rio Grande City a maximum temperature of 105° has been recorded for the months of April and June, and at Fort Stockton 106° in June. At Brownsville the highest temperature is reached in June. Fort Elliott, in the northwestern part, or panhandle, of the State, enjoys the coolest summers, the thermometer for the three months not rising above 86.7°.

One of the prominent features of the climate of Texas is what is commonly known as the "Texas Norther." Its coming marks a sudden and extreme change of temperature, produced by a rush of cold wind from the North, usually appearing unannounced, though occasionally indicated by a peculiar haze in the Northern sky. The northers are usually preceded by a warm spell of twenty-four hours or so of extremely high temperature, and the change of temperature within a few minutes is very great, sometimes in the winter months falling as much as 30° or 40°, though usually much less. The northers continue about three days, the second day usually being the coldest, and they are invariably followed by warm weather, though at times the northers follow each other so closely as to produce eight or ten days of cold. They may be expected at all times of the year, and it is customary for travelers to be provided with blankets, even for a trip of a few days. The northers are sometimes accompanied by rain, and this phenomenon being peculiar to Texas, they are classed as dry or wet northers. In summer the northers are not so frequent as in winter, nor are they marked with so great extremes of temperature. In summer also there are warm as well as cold northers.

The elevation of Western Texas is from 400 to 1000 feet above the sea, and there the temperature is dry, very invigorating and entirely free from fogs or malaria. The climate is about that of the highlands of Mexico. The climate there receives some of its mildness from the great ocean current or Gulf stream of the Atlantic, which makes its circuit of 10,000 miles, bringing its heat from the equatorial region and throwing its warm stream hundreds of miles inland, and it fortunately escapes the chilly winds of the Florida coast, caused by the cold water coming from the North and insinuating itself between the land and the Gulf stream, the coast of Western Texas being hundreds of miles beyond its terminus. Western Texas is again favored by nature in the abundance of the ozone in the atmosphere. This element is so abundant that meats are perfectly preserved in the open air without being previously salted, the bodies of hundreds and thousands of animals lying dead on the prairie emitting no odor whatever. It is this, with the other elements of pure air, that renders Western Texas one of the few places on the surface of the globe where the consumptive can dwell in comparative comfort and add a few years to the time allotted him in a less favorable climate. It is a well-established fact that yellow fever cannot prevail in Western Texas as an epidemic.

With regard to the seasons the winter months seem to be the dryest, the precipitation throughout the State varying from three to seven inches, and even less than that in Denison county. During the spring months the rainfall is greatest in the Eastern counties, amounting to from twelve to thirteen inches. During the summer months the country around Corsicana has suffered much from droughts, though some years there is an abundant waterfall there. The maximum for the whole State for the year is about twenty-one inches. The fall months differ little from those of summer, except that there is more rain in the Eastern counties. From the reports given, San Antonio seems to enjoy the greatest regularity in its monthly rainfalls, there being but one month in the year when it is less than two inches, while its maximum for any month of the year is 8.6 inches. The record of Corsicana shows very nearly the same regularity, a maximum of 7.7 inches.

The timbered region of Eastern Texas is better supplied with water than any other part of the State. Springs of good freestone water are found in almost every county, and wells furnish an abundant supply for domestic purposes. The small streams usually become dry during the summer months, and artificial reservoirs or simple earth embankments collect a sufficient supply of water during the rainy season for farm and stock purposes.

The geographer of the Census Bureau in the last reports gives the following description and classification of the topography and soils of Texas:

"The State of Texas with its immense territory naturally presents agricultural features greater in variety, perhaps, than those of any other State in the Union. Its position at the southwestern extremity of the agricultural regions of the South gives to a part of the State features similar in most respects to other Southern States. Including as it does the southeastern borders of the great Western plains, the lands of the western part of the State resemble those of New Mexico. Those of the gypsum formation and of the red loam region seem only to extend northward into the Indian Territory and Oklahoma. The following agricultural regions may be conveniently distinguished:

1. Timbered Upland Region of East and Central Texas.
2. Southern and Coast Prairie.
3. Central Black Prairie.
4. Northwestern Red Loam Lands.
5. Western and Northwestern Uninhabited Region.
6. River Alluvial Lands, including the Brazos Delta.

In the matter of agricultural products Texas stands at the head of the cotton-producing States, the crop in 1899 being 1,708,000 bales, as against 310,000 of Georgia, the next largest cotton-producing State. In sugar Texas is the second State, Louisiana being first, the product in 1899 being for Louisiana 597,963,187 pounds, and for Texas 11,882,852. The corn crop in recent years has represented an acreage of upwards of 3,000,000, and the annual product is about 72,-000,000 bushels. Texas stands fifth among the corn-producing States. While the wheat crop of Texas is large, it is far less important than many of its other crops. Last year the product was upwards of 7,000,000 bushels. The acreage of oats last year was 652,446, and the product aggregated 16,311,000 bushels.

Galveston is the natural port of Texas. Thither the railroads from Houston go by the shortest route to the seaboard. Aransas Pass, Sabine Pass and Corpus Christi each fought hard for the blessing which fell to the Island City. Its location with reference to the productive resources of Texas settled the matter for all time to come. No matter if Galveston is leveled by the waters of the Gulf in another fifty or one hundred years, it will yet rise again and remain then as now the chief port of Texas. The forces inland and on the sea which caused the railroad builders to choose Galveston as the great shipping point of the State are still operative. As well talk of abandoning Chicago after the fire which swept away her wealth in a single day as to talk of abandoning Galveston after the late disaster. The reasons that have made Galveston great in the past are those which settled the fate of New York, Philadelphia, Boston, Chicago, London and Canton; that is, facility of communication with the interior and a good harbor for the outgoing commerce. Galveston will be a great city in the years to come, even if it becomes necessary to fill the island with earth to lift it above the floods. Galveston has within her grasp the terminals of the transportation lines of Texas and the Southwest, and no other port on the Gulf of Mexico has, and there the argument ends in favor of Galveston.